# Android App开发者必修16堂课

赵令文 Brad 著

化学工业出版社

·北京·

本书用轻松易懂的语言和生动活泼的形式,介绍了Android App的开发技巧,主要内容包括:开发环境建置与基本使用、基本程序运行原理与应用、基本用户界面与事件触发、对话框与通知事件处理、进阶程序运行原理与应用、菜单与动作列处理、自定义View与Surface View、数据存取、因特网相关、影音多媒体与相机、地图与卫星定位系统、感应器运行原理及应用、资源与国际化、系统功能与装置控制、实际项目开发、App发布等。

本书内容起点低、容易上手、范例经典、源代码步步解析,图文并茂,帮助初学者快速入门!本书非常适合Android App开发初学者、技术人员以及业余爱好者阅读使用。

### 图书在版编目(CIP)数据

Android App开发者必修16堂课/赵令文著. —北京:化学工业出版社,2015.8

ISBN 978-7-122-24412-3

Ⅰ.①A… Ⅱ.①赵… Ⅲ.①移动终端–应用程度–程序设计 Ⅳ.①TN929.53

中国版本图书馆CIP数据核字(2015)第138932号

原繁体版书名:Android App開發者必修16堂課　作者:趙令文　Brad

ISBN 978-986-199-398-0

本书中文繁体字版本由城邦文化事业股份有限公司电脑人文化在台湾地区出版,今授权化学工业出版社在中国大陆地区出版其中文简体字平装本版本。该出版权受法律保护,未经书面同意,任何机构与个人不得以任何形式进行复制、转载。

项目合作:锐拓传媒copyright@rightol.com

北京市版权局著作合同登记号:01-2016-0263

---

责任编辑:李军亮　耍利娜　　　　　　文字编辑:吴开亮
责任校对:边　涛　　　　　　　　　　装帧设计:刘丽华

---

出版发行:化学工业出版社(北京市东城区青年湖南街13号　邮政编码100011)
印　　刷:北京永鑫印刷有限责任公司
装　　订:三河市宇新装订厂
710 mm×1000mm　1/16　印张25$\frac{1}{2}$　字数527千字　2016年2月北京第1版第1次印刷

购书咨询:010-64518888(传真:010-64519686)　售后服务:010-64518899
网　　址:http://www.cip.com.cn

凡购买本书,如有缺损质量问题,本社销售中心负责调换。

---

定　价:88.00元　　　　　　　　　　　　　　　版权所有　违者必究

# 前　言

笔者从事Android App开发的实务教学已经三年多了，这三年多来的变化也相当多。不只是开发工具或是API上的变化，还有App市场上的变化。现在的您还在玩Angry Bird吗？还是开始换玩Candy Crush了？之前Google play在台湾地区消费市场上的风风雨雨，终于又可以开始贩卖购买付费App了，三年前相关单位办了一场"一千五百万创意成金"活动，只要提案通过审核并如期上架，一个App就可能获得3万~8万元奖金（敝人也提了三个App获得奖金），而现在的市场机制没有这样的活动了，转变为鼓励的是质量较佳的App。敝人也从教育训练市场间接看到就业市场上的需求不断攀升。

很多初学者想要快速学习Android App的开发，却往往适得其反，虽然相关学习资料很多，但是大部分都是片段的技术数据。可能很容易找到如何发出通知消息的知识，却不知该如何应用或是做出不同的变化；找到数据并将程序代码复制后可以反转手机变成静音，不知如何运用手机传感器成为方向控制器等。相信我，你绝对可以找到成千上万的Android App开发秘诀，但是也许并不适合初学者的学习。除非你知道每一行程序代码的存在意义，否则复制粘贴绝对是最糟糕的学习模式。我的上课模式不会有幻灯片，也没有事先写好的程序代码，就是从项目建立开始开发，写出来的每一行代码都要清楚在做什么，观念的建立非常重要，观念清楚之后，想要做出不同的变化就不成问题了。因此，本书希望可以将初学者导向建立观念式学习模式，不是提供片断的秘诀而已。书中最后专题范例就是2012年在"资策会"移动装置开发班的Android游戏开发课程中以12小时授课时间，从项目建立开始实际开发，再利用课后时间进行修正微调后的作品，获得了2012年某电信社会组优胜作品的殊荣。

我不是专家，只是爱玩而已。2011年以改编Lode Runner经典游戏参加其他比赛，专家评审建议加上自编关卡网络分享功能，可以使这个App有加分效果。当下敝人极度不同意，因为这样的行为表面看似增加玩家之间的互动，事实上却增加玩家游戏的潜藏隐私安全性风险。最后为了继续参赛而将该功能开发上去，并在电视上展现实时分享功能。赛后自行另外以原先版本约晚三个月在Google play上架发行。一年半过去，单纯游戏版本的实际安装数量已经是网络分享版本数量的两倍以上。

　　最后以此书献给最挚爱的家父赵光明先生与家母谢宝秀女士，他们奉献毕生心力于教育事业。投身教育训练的敝人正是受到他们不断的鼓励指导，才能完成此拙著。当然，也要感谢陪伴我赶稿的老婆，及帮我测试玩App游戏的孩子们。另外，更是感谢协助处理App视觉艺术部分的墨比斯－云云手。

赵令文Brad<br>
2013.5.4

# 目 录

## 第1课　开发环境建置与基本使用 ······························································ 1

1-1　学习开发的基本概念 ······················································································ 2
  1-1-1　Java语言的角色 ················································································ 2
  1-1-2　Unix/Linux的文件系统 ······································································ 2
  1-1-3　学习目标 ··························································································· 3
1-2　安装JDK ········································································································· 3
1-3　安装Eclipse ···································································································· 4
1-4　设定Eclipse ···································································································· 5
1-5　安装设定Android SDK ···················································································· 8
  1-5-1　在Eclipse外挂ADT ············································································ 8
  1-5-2　建立及使用仿真器 ·············································································10

## 第2课　基本程序运行原理与应用 ············································· 13

2-1　"Hello, World？Hello, Lottery！" ···································································14
  2-1-1　建立新项目 ·······················································································14
  2-1-2　版面配置 ··························································································17
  2-1-3　开发程序 ··························································································20
  2-1-4　安装执行测试 ···················································································22
2-2　"BMI？Lottery！" ··························································································23
  2-2-1　存取控制元件 ···················································································23
  2-2-2　按钮事件处理模式 ············································································24
  2-2-3　开发设计功能 ···················································································26
  2-2-4　修改程序 ··························································································27
2-3　写完了，然后呢? ···························································································28
  2-3-1　加上欢迎界面 ···················································································28
  2-3-2　调整启动程序 ···················································································31
2-4　Activity的生命周期 ·······················································································34

  2-4-1 生命周期的观念 …………………………………… 34
  2-4-2 测试实作 ………………………………………… 36
  2-4-3 开始观察 ………………………………………… 39
 2-5 Activity 切换 Activity ……………………………………… 40
  2-5-1 仅作启动切换 …………………………………… 40
  2-5-2 传递数据过去 …………………………………… 40
  2-5-3 切换之后回来确认 ……………………………… 41
  2-5-4 将数据传递回来 ………………………………… 42
 2-6 Service 的运行应用 ………………………………………… 44
  2-6-1 生命周期实测 …………………………………… 44
  2-6-2 与线程共舞 ……………………………………… 48
  2-6-3 通过 Broadcast 发送数据给前台 ……………… 49

# 第 3 课 基本用户界面与事件触发 ……………………………… 53

 3-1 条列显示元件 ListView ……………………………………… 54
  3-1-1 基本格式 ………………………………………… 54
  3-1-2 进阶格式 ………………………………………… 57
 3-2 线性配置 LinearLayout …………………………………… 59
 3-3 相对配置 RelativeLayout ………………………………… 63
 3-4 表格配置 TableLayout …………………………………… 68
 3-5 网格显示 GridView ………………………………………… 71
 3-6 滑动显示 ViewFlipper ……………………………………… 75

# 第 4 课 对话框与通知事件处理 ………………………………… 83

 4-1 AlertDialog 对话框的使用 ………………………………… 84
  4-1-1 建立 AlertDialog 对象 ………………………… 84
  4-1-2 消息对话框 ……………………………………… 84
  4-1-3 确认对话框 ……………………………………… 87
  4-1-4 选择式对话框 …………………………………… 89
  4-1-5 进阶选择式对话框 ……………………………… 92
 4-2 自定义对话框（Dialog）与日期时间对话框 …………… 95
  4-2-1 自定义对话框 …………………………………… 95
  4-2-2 日期选择对话框 ………………………………… 99

|  |  |  |
|---|---|---|
| 4-2-3 | 时间选择对话框 | 101 |
| 4-3 | Toast 及自定义 Toast | 103 |
| 4-3-1 | 一般的 Toast | 103 |
| 4-3-2 | 自定义 Toast | 104 |
| 4-4 | 进度显示对话框 | 107 |
| 4-5 | 通知列处理模式 | 110 |
| 4-5-1 | 版本差异 | 110 |
| 4-5-2 | API Level 11 之前 | 111 |
| 4-5-3 | API Level 11+ | 111 |
| 4-5-4 | 应用场合 | 114 |

## 第 5 课　进阶程序运行原理与应用　　115

|  |  |  |
|---|---|---|
| 5-1 | 多重线程 Thread | 116 |
| 5-1-1 | 开发重点观念 | 116 |
| 5-1-2 | 存取 View 组件 | 119 |
| 5-1-3 | 提早结束线程的生命周期 | 120 |
| 5-1-4 | 另外一种开发方式 | 121 |
| 5-2 | 定时及周期任务（Timer & TimerTask） | 123 |
| 5-3 | 同步任务 AsyncTask | 126 |
| 5-3-1 | 使用观念 | 126 |
| 5-3-2 | 生命周期 | 126 |
| 5-3-3 | 定义泛型参数 | 128 |
| 5-3-4 | 基本开发程序 | 129 |
| 5-3-5 | 程序架构 | 129 |
| 5-4 | 倒数定时器 | 133 |
| 5-4-1 | 开发模式 | 133 |
| 5-4-2 | 直接实作练习 | 133 |

## 第 6 课　菜单与动作列处理　　137

|  |  |  |
|---|---|---|
| 6-1 | 菜单 Menu | 138 |
| 6-1-1 | Options menu 选项菜单（硬件菜单键） | 138 |
| 6-1-2 | Context menu 内容菜单 | 141 |

6-1-3　Popup menu 弹出式菜单 145
6-2　动作列 Action Bar 147

# 第7课　自定义 View 与 SurfaceView … 155

7-1　自定义 View：继承 View 156
7-2　自定义 View 与触控手势事件处理 165
　　7-2-1　一般触控事件侦测处理 165
　　7-2-2　手势侦测事件处理 166
7-3　自定义 SurfaceView：继承 SurfaceView 170
7-4　以自定义 View 来实现手写签名 App 范例实作 174
　　7-4-1　前期准备 175
　　7-4-2　开始处理签名的手势侦测处理 177
　　7-4-3　处理外部功能 182

# 第8课　数据存取 … 185

8-1　偏好设定 186
　　8-1-1　处理方式 186
　　8-1-2　基本处理程序 186
　　8-1-3　范例说明 186
　　8-1-4　完整范例 189
8-2　内部文件存取机制 191
　　8-2-1　使用观念 191
　　8-2-2　写出基本程序 191
　　8-2-3　读入基本程序 193
8-3　外部文件存取 195
　　8-3-1　SDCard 文件系统基本概念 195
　　8-3-2　判断 SDCard 的挂载点（Mount Point） 196
　　8-3-3　应用程序文件应该在哪里 196
　　8-3-4　开启写出数据的权限 196
　　8-3-5　开始进行程序开发 198
　　8-3-6　写出数据文件 198
　　8-3-7　读入数据文件 199

- 8-4 移动装置数据库处理机制 SQLite ………………………………… 200
  - 8-4-1 建立数据库的辅助类别对象 ………………………… 200
  - 8-4-2 预先处理模式 ………………………………………… 200
  - 8-4-3 简单查询数据 ………………………………………… 202
  - 8-4-4 新增数据 ………………………………………………… 203
  - 8-4-5 删除数据 ………………………………………………… 203
  - 8-4-6 修改数据 ………………………………………………… 204
  - 8-4-7 进一步了解查询方式 ………………………………… 204
- 8-5 应用 App 资源中的数据存取数据：以游戏关卡数据处理为例 …… 205
  - 8-5-1 定义数据 ………………………………………………… 206
  - 8-5-2 读取数据文件 …………………………………………… 207
  - 8-5-3 程序中读取方式 ………………………………………… 207

# 第 9 课　因特网相关 ……………………………………………… 209

- 9-1 网络接口及 IP Address ………………………………………… 210
  - 9-1-1 装置的网络状态 ………………………………………… 210
  - 9-1-2 网络接口的 IP Address ………………………………… 210
  - 9-1-3 取得装置联机 IP Address ……………………………… 212
  - 9-1-4 建构 IP Address 对象实体 …………………………… 213
- 9-2 UDP 通信协议的数据存取 ……………………………………… 214
  - 9-2-1 处理模式 ………………………………………………… 214
  - 9-2-2 实作测试 ………………………………………………… 214
- 9-3 TCP 通信协议的数据存取 ……………………………………… 220
  - 9-3-1 处理模式 ………………………………………………… 220
  - 9-3-2 实作测试 ………………………………………………… 220
- 9-4 Http 通信协议的数据存取 ……………………………………… 225
  - 9-4-1 以 AndroidHttpClient 及 DefaultHttpClient 实作 …… 225
  - 9-4-2 以 java.net.HttpURLConnection 实作 ………………… 228
- 9-5 WebView 使用 …………………………………………………… 229
  - 9-5-1 基本的处理方式——直接放进 Activity 中 …………… 229
  - 9-5-2 基本的处理方式——以版面配置方式处理 …………… 230
  - 9-5-3 进一步设定 WebView 功能 …………………………… 236

## 第10课　影音多媒体与相机　　243

### 10-1　播放音乐　　244
- 10-1-1　基本概念　　244
- 10-1-2　SDCard 上的音乐播放　　245
- 10-1-3　播放项目资源中音乐文件　　247
- 10-1-4　播放 URL 的音乐文件　　247
- 10-1-5　暂停继续播放　　248
- 10-1-6　停止播放　　248

### 10-2　音效处理　　249
- 10-2-1　建构 SoundPool 对象实体　　249
- 10-2-2　实时播放音效　　250

### 10-3　录音处理　　250
- 10-3-1　调用其他录音程序　　251
- 10-3-2　自定义录音处理程序　　252

### 10-4　录像放映　　254
- 10-4-1　录像　　254
- 10-4-2　调用其他录像程序　　254
- 10-4-3　自定义录像程序　　256
- 10-4-4　播放影片　　258

### 10-5　相机　　259
- 10-5-1　调用其他照相程序　　259
- 10-5-2　自定义相机程序　　261

## 第11课　地图与卫星定位系统　　267

### 11-1　GPS 定位　　268
- 11-1-1　开始基本实作　　268
- 11-1-2　较佳位置取得　　270

### 11-2　基本 Google Map　　275
- 11-2-1　开发前期作业　　276
- 11-2-2　Hello，Map　　277
- 11-2-3　在 Android 上开发的应用　　279

### 11-3　进阶 Google Map　　280

| 11-3-1 | JavaScript 处理说明 | 280 |
| 11-3-2 | JavaScript 数据传回 Android | 282 |
| 11-3-3 | 以 Android 传递数据给 JavaScript | 283 |

## 第12课　传感器运行原理及应用 …………………………… 285

- 12-1　传感器运行原理与应用　286
  - 12-1-1　基本概念　286
  - 12-1-2　处理原则　286
  - 12-1-3　实作开发　287
  - 12-1-4　用户装置支持处理　288
- 12-2　三轴加速传感器　289
- 12-3　重力加速度传感器　293
- 12-4　磁极方向传感器　296
- 12-5　光线/温度/湿度/压力传感器　300

## 第13课　资源与国际化 ……………………………………… 305

- 13-1　提供资源内容　307
  - 13-1-1　预设资源内容及架构　307
  - 13-1-2　替代选择性资源内容　309
- 13-2　存取资源内容　311
  - 13-2-1　程序代码中存取资源内容　312
  - 13-2-2　XML 中存取资源内容　313
- 13-3　应用程序执行中的改变　314
  - 设计一个保留及回存对象　314
- 13-4　资源内容的区域化　314
  - 13-4-1　支持的区域国别（地区）　315
  - 13-4-2　进一步认识项目资源　316
  - 13-4-3　资源类型　322
  - 13-4-4　区域化确认检查　323

## 第14课　系统功能与装置控制 ……………………………… 325

- 14-1　移动装置相关辨识　326

14-2 移动电话通话状态 328
14-3 移动电话用户相关数据 330
　14-3-1 用户账号 330
　14-3-2 取得联系人姓名 331
　14-3-3 用户的相簿 332
14-4 开发者基本道德 332

## 第15课　实际项目开发　335

15-1 弹指砖块王（Bricks Fighter） 336
　15-1-1 App简易架构 337
　15-1-2 欢迎页面 337
　15-1-3 游戏关卡菜单 339
　15-1-4 游戏主页 344
15-2 掏金沙（Lode Runner） 353
　15-2-1 开发动机 353
　15-2-2 着手规划 354
　15-2-3 游戏架构 355
　15-2-4 关卡菜单 358
　15-2-5 游戏画面 359
　15-2-6 关卡地图 362
　15-2-7 敏感争议 371
15-3 炸弹超人（Bomb King） 371
15-4 其他应用程序开发项目 374
　15-4-1 个性签名产生器 374
　15-4-2 开发观念原则 386

## 第16课　App发布　387

16-1 包装发布到Google Play 388
　16-1-1 包装成为APK 388
　16-1-2 首次注册开发者 390
　16-1-3 发布APK到Google Play 392
16-2 App创意开发与比赛经验心得分享 393

# 01 Chapter

## 第1课 开发环境建置与基本使用

1-1 学习开发的基本概念

1-2 安装JDK

1-3 安装Eclipse

1-4 设定Eclipse

1-5 安装设定Android SDK

## 1-1 学习开发的基本概念

Android App的开发应用在这几年之间已经非常普遍。光是Google play上面的App数量就非常惊人，还不包含其他各家市场，以及各不同行业领域业者提供自家相关的App。事实上，Android App的开发应用上手非常容易，只需要基本的Java程序设计能力，加上熟练使用Android API即可开始进行开发，而本书的重点就是放在Android API的学习上。

### ■ 1-1-1 Java语言的角色

Android SDK的开发模式完全就是采用Java程序语言。所以开始开发之前，最好能够对Java语言熟悉了解，这样绝对有助于开发过程。其实有许多刚开始入门的学习者，可能一心想要快速学习开发Android App，于是就买书、上网学习，当看到与自己打算开发项目类似功能的范例时，就马上进行复制粘贴，顺利的话，可以先看到原开发者呈现的执行效果；但是大部分却都不是那么顺利，可能是API Level设定下载不同，可能是开发环境或是仿真器差异，或是编译使用函数库放置路径不同等一系列问题，还可能不是解决一个问题就结束。因此就需要开发Java语言的基本观念来进行排解，所以Java的基本认识相当重要。

当可以开始将原开发者的原始码执行之后，想要修改成为自己想要的模式。此时最基本的动作就是认识原开发者当时设定各自变量所代表的意义，才能正确地进行修改调整，而一般开发者的变量名称的命名应该都还容易判断出来，但是整段程序方法的作用是什么？应该如何下手呢？还是需要Java语言的基本观念来处理。

那么就不要复制粘贴的学习模式，从头开始来学习开发，这就是完全在写Java程序语言。

### ■ 1-1-2 Unix/Linux的文件系统

Android操作系统来自于Linux，所以其文件系统是以Linux的单个操作系统来进行的。如果是熟悉MS Windows的多磁盘的操作系统的学习者，只需要想成只有一个C磁盘驱动器而已，既然只有一个C磁盘驱动器，就干脆不需要提及C磁盘驱动器这件事。再来就是路径符号刚好与MS Windows的操作系统相反，而是使用"/"反斜线，对于许多程序语言开发而言，"\"斜线符号是和跳脱字符一样，容易造成开发上的困扰，例如在使用表示网址的路径符号时也是使用"/"反斜线（大部分的操作系统都是使用"/"反斜线，少部分的操作系统是使用"\"斜线）。再来就是大小写严格区分这件事与MS Windows是不一样的。

大致上常见的开发上差异如此而已，也不需要先去熟悉了解Linux之后再来学习Android。

### ■ 1-1-3　学习目标

既然想学习Android App开发，总要有个想要开发的目标。有个目标想要开发，学习效果就会比较好；没有特定目标的话，可以依照本书上面的项目来进行。

本书无法提供创意的思考模式，如果硬要说不在本书设定的主题范畴内，那是借口。因为笔者本身没有这方面的专长，所有笔者开发的项目都是来自于个人的需求所产生的。因为偏爱当年红白机的游戏，所以想自行改编开发Lode Runner，炸弹超人等；因为从小第一次接触的电玩是"打砖块"，所以也来改写回味一番；因为自己想要在平板计算机上面有个好用的万用笔记功能，可以有一般笔记，同时照相录音录像，卫星定位等多样化的功能，所以就想要自行开发，这就是我的创意来源。

> 重点：复制粘贴是最差的学习模式，可能一开始会有感觉（错觉），中间一定会卡，而且卡很久。只要是写在自己程序中的代码，就不能有不认识的东西，开发出来的App才是你的。

## 1-2　安装JDK

Java语言是Android的基本，建议先将开发环境安装上JDK（Java Development Kit），这是当年Sun公司针对Java程序语言的开发人员发行的SDK（Software Development Kit），自从2006年之后，Sun公司宣布基于GPL协议使其成为自由软件。

下载链接http://www.oracle.com/technetwork/java/javase/downloads/ index.html
如下画面：

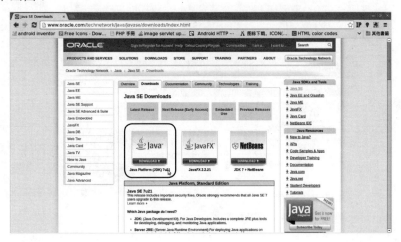

单击"Java Platform（JDK）"进入如下画面。

就可以依照开发操作系统平台，下载适当的版本进行安装程序。

## 1-3 安装Eclipse

Eclipse是开发Android App的主要整合开发工具。所谓的"整合开发工具"，就是在开发过程中，从基本的程序编辑器开始，除错工具、Log记录、仿真器等全部具备。开发者开始执行Eclipse之后，所需要的开发资源就大部分都具备了，是相当方便的开发利器。

先至官方网站http://www.eclipse.org，找到Download Eclipse链接过去，看到如下网页。

找到"Eclipse IDE for Java Developer"项目下载。下载下来的文件只需要解压缩到自己喜欢的特定目录下即可，无须安装程序。

就在解压缩后的目录下找到"eclipse"，然后运行。

当询问使用的工作区路径的时候，此时只是决定这次执行预设使用的工作区路径，可以另外指定或是沿用询问的默认值，无论如何，重点是要知道开发的项目放在什么路径下即可。

## 1-4 设定Eclipse

首次来到Eclipse的整合开发环境，看到以下欢迎的画面。

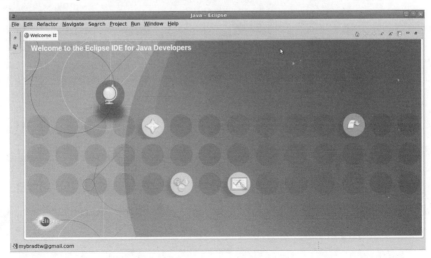

直接将该页关闭，看到以下可能会令首次接触Eclipse的开发者不知所措的画面。

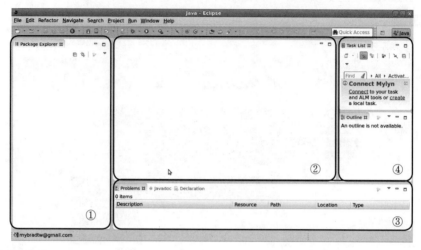

简要说明：

① 左侧有个Package Explorer区，用来显示目前工作区下管辖的项目列表，及处理项目架构下的文件目录，非常重要。

② 中间空白区域就是平常开发程序的编辑器。

③ 下方有其他辅助开发的视图，以个别页签进行切换。

④ 右侧两个暂时不需要，可以直接关闭。

以下简略介绍基本的设定项目。

① 设定程序编辑器字体放在上方菜单列"Window"→"Preference"。

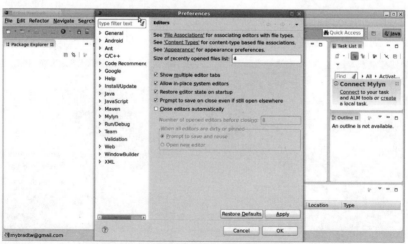

② 左侧内容视安装外挂而定，找到Gereral后单击展开，"Apperance"→"Colors and Fonts"，看到中间窗口的Basic展开。

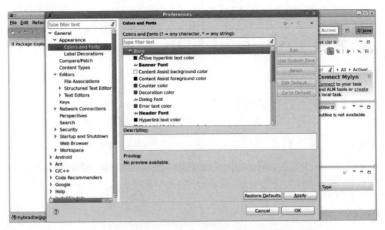

③ 找到"Text Font"→"Edit..."就可以设定编辑器的字体。

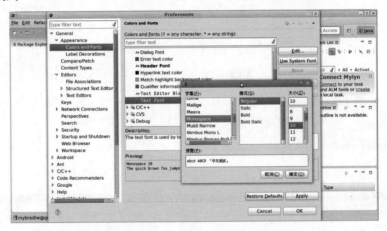

④ 再来设定编辑程序代码的文字编码,展开"General"→"Workspace",建议读者使用UTF-8。

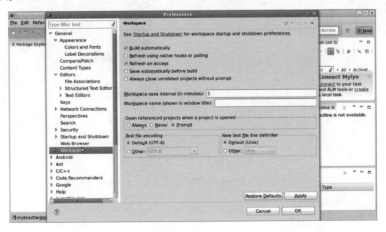

## 1-5 安装设定Android SDK

### ■ 1-5-1 在Eclipse外挂ADT

先到Android开发者官方网站 http://developer.android.com

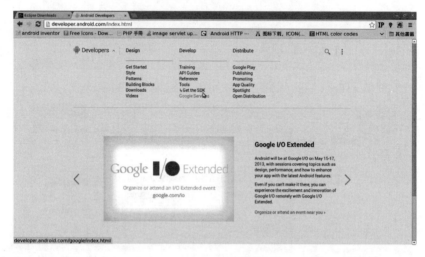

点选"Design"→"Tools"→"Get the SDK"后，看到如下网页。

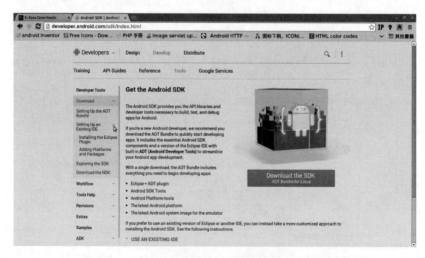

在左侧的"Setting Up an Existing IDE"下的"Installing the Eclipse Plugin"后看到如下网页内容。

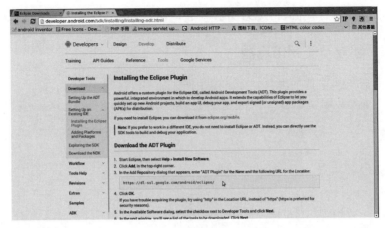

将其网页内提及Download the ADT Plugin中的网址：https://dl-ssl.google.com/android/eclipse/

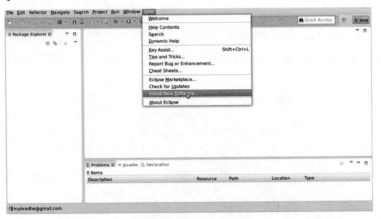

复制下来，回到Eclipse中，在菜单列中"Help"→"Install New Software..."。

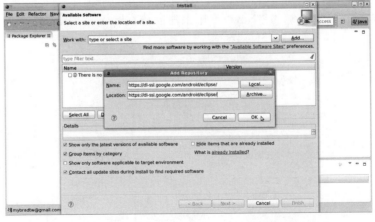

出现的对话框中，按下"Add"后，将刚才复制的网址粘贴上。

这样就开始进行Android ADT外挂安装，相当容易。

## ■ 1-5-2　建立及使用仿真器

安装之后，在工具列中找到如下图中所示。

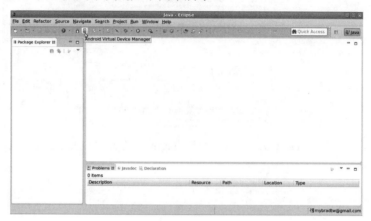

Android Virtual Device Manager用来进行仿真器的建立与启动相关管理工具，按下之后如下所示。

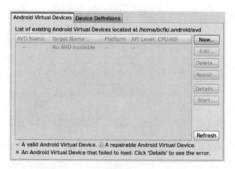

接着按下对话框右侧的"New..."用来建立仿真器。

在以上对话框中输入想要建立仿真器的基本数据。
- AVD Name：自定义名称。
- Device：设定想要仿真的装置。
- Target：安装 Android 的版本。
- Keyboard：务必勾选 Hardware keyboard present。
- Skin：务必勾选 Display a skin with hardware controls。
- Front Camera：如果有安装 Cam 的读者可以考虑使用。
- SD Card：请适量输入模拟 SD Card 的空间大小。

按下"OK"后启动看看（可能会需要稍微久的时间进行启动，请耐心等候）。

按下屏幕上的"OK"，出现如下画面。

其他部分就不一一赘述，接下来就可以开始玩这只手机啰。

# MEMO

# 02 Chapter

# 第2课 基本程序运行原理与应用

2-1 "Hello，World？ Hello，Lottery！"

2-2 "BMI？ Lottery！"

2-3 写完了，然后呢?

2-4 Activity的生命周期

2-5 Activity切换Activity

2-6 Service的运行应用

## 2-1 "Hello，World? Hello，Lottery！"

关于"Hello，World"项目的开发，相信读者可以在100本的Android相关书籍中看到。因此，笔者不打算浪费篇幅来介绍，直接来写一个"Hello，Lottery！"，就是利用计算机选号来产生乐透号码的程序，程序开发逻辑上非常简单，重点放在整个开发项目架构，以及一个App的执行生命周期的相关学习。事实上，读者只需要略作些许修改，就可以放在Google play上面发布，而目前约有101个乐透号码产生器的App在Google play上面赚取广告费。

### ■ 2-1-1 建立新项目

先进行项目的建立，点选窗口菜单列的选项"File"→"New"，显示子菜单中如果出现"Android Application Project"，那就点选下去；但是大部分一开始建立开发环境之后，应该还不会出现在这个子菜单，所以请在这个子菜单中点选"Other..."选项，之后将会出现一个对话框，如下图所示。

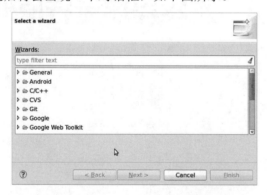

实际画面不一定与笔者的相同，其内容会因Eclipse中所外挂安装的软件而有所不同，重点是找出Android的文件夹，并将其点选展开后如下图所示，点选"Android Application Project"，并按下下方的"Next>"按钮。

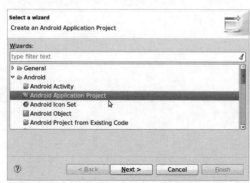

正式开始新项目的建立，首先针对新项目的名称来处理。有以下几个重要项目填写。

• Application Name（应用程序名称）：也就是用户看到的名称。

• Project Name（项目名称）：是开发者自行定义的名称，通常会与 Application Name 相同，但是为了使读者了解两者的差异性，故意写成两个不同的内容。

• Package Name（软件名称）：是这套软件安装在用户装置下的识别名称。用户可能会在其移动装置上装了来自于世界各地开发者的软件，而这些软件是以 Package Name 识别方式来存放这些不同的开发软件的。通常在 Java 程序项目开发中，惯例是以开发者的公司组织学校，或是个人的专属域名倒过来命名。例如笔者有一个开发测试域名为 ez2test.com，则开发 Android 的游戏类的"炸弹王（BombKing）"软件，可能在此处就是 com.ez2test.android.games.bombking。因此，目前的"Hello，Lottery"项目数据如下。

• Application Name：亿万富翁大乐透。

• Project Name：HelloLottery。

• Package Name：tw.brad.android.book.hellolottery。

• 其他下面的四个项目，就先以默认值为主，先不进行处理。

这些项目都可以在项目建立之后再做修改调整，包含上述的三个命名项目。如下图所示。

按下"Next>"按钮之后的对话框就都可以直接按下"Next>"按钮跳过，一直到可以按下"Finish"，新项目就建立完成。成果如下列图示程序所示。

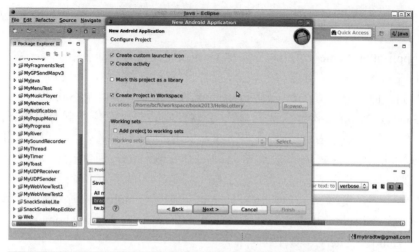

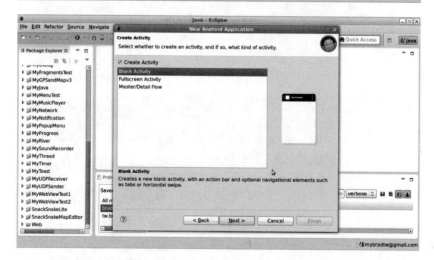

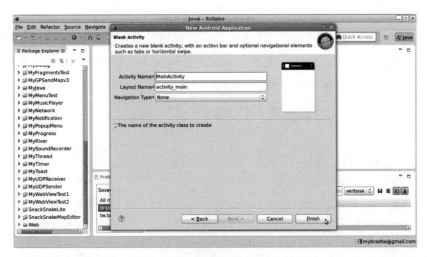

当按下"Finish"按钮之后,将会开始自动产生新项目的架构,可能会需要几秒时间,请耐心等候,千万不要以为死机或是其他不可预期状况发生。

等到以下画面出现。

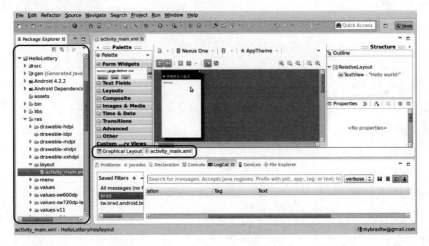

左边就是项目目录架构,Project Name 就出现在此,以下介绍一开始必须先认识的子目录。

- src/:开发的程序。
- res/:应用程序使用到的相关资源。
- drawable-xxx/:应用程序项目的影像图形文件。
- layout/:应用程序的版面配置文件,就是呈现的外观部分。
- values/:应用程序项目使用的数据值。

### ■ 2-1-2 版面配置

目前就是停留在版面配置文件 activity_main.xml 上面,使得右边开发视窗要呈现该文件的内容，可以在该窗口下方看到两个页签。
- Graphical Layout：以可视化方式来规划配置版面（笔者建议用来参考观看用）。
- activity_main：以XML文件格式来规划配置版面（笔者强烈建议开发使用）。

因此，请点按下"activity_main"的页签，将会出现如下XML文件格式，内容如下：

```xml
<RelativeLayout xmlns:android="http://schemas.android.com/apk/res/android"
    xmlns:tools="http://schemas.android.com/tools"
    android:layout_width="match_parent"
    android:layout_height="match_parent"
    android:paddingBottom="@dimen/activity_vertical_margin"
    android:paddingLeft="@dimen/activity_horizontal_margin"
    android:paddingRight="@dimen/activity_horizontal_margin"
    android:paddingTop="@dimen/activity_vertical_margin"
    tools:context=".MainActivity" >

    <TextView
        android:layout_width="wrap_content"
        android:layout_height="wrap_content"
        android:text="@string/hello_world" />

</RelativeLayout>
```

以上内容较复杂，先以架构来分析：
- <RelativeLayout>
- <TextView />
- </RelativeLayout>

最外层定义了一种版面配置方式，其名称为RelativeLayout（其实就是Java的类别名称，通常有个命名特性，就是驼峰式命名，首字母大写，其他小写，如果是复合字，就很像骆驼的驼峰）。该类别对象具有容器特性，也就是说可以装进其他显示组件，因此，会另外有相对应的结尾标签</RelativeLayout>，前置斜线符号字符，也就是有头有尾的完整结构。而目前装载了一个TextView的显示组件，用来显示一般的文字内容，并非容器特性的显示组件，虽然也可以有头有尾地加上</TextView>，但是通常为了避免麻烦，就直接在最后加上斜线符号字符<Xxx/>的方式，有头无尾自我结束。

而其中会有一些android:xxx="xxx"的项目，称其为属性设定。等号左边为属性项目，等号右边为设定值，例如：

```xml
android:layout_width="wrap_content"
```

表示属性设定 android:layouy_width 该组件在版面宽度上，设定为"wrap_context"依照内容决定其显示宽度。注意一点就是设定值一定被包在双引号之中，因为 XML 是文件格式，所有数据都是字符串形式。

至此，先简单修改如下：

```
<LinearLayout xmlns:android="http://schemas.android.com/apk/res/android"
    android:layout_width="match_parent"
    android:layout_height="match_parent"
    android:orientation="vertical"
>

<Button
    android:id="@+id/torich"
    android:layout_width="match_parent"
    android:layout_height="wrap_content"
    android:text="致富按钮" />

<TextView
    android:id="@+id/richnum"
    android:layout_width="match_parent"
    android:layout_height="wrap_content"
    android:gravity="center_vertical|center_horizontal"
    android:text="此处将会出现致富号码..." />

</LinearLayout>
```

往下阅读学习过程中，请记得笔者建议的口诀，无论是本书或是其他文章学习，请不要将不认识的内容复制粘贴到自己的开发项目中，那可能会造成处理上不必要的麻烦，甚至于错误。所有自行开发的程序内容都应是自己可以掌握的东西。

所以，以下先就目前简单的内容作说明。

• LinearLayout 是一种版面配置，算是最简单的版面配置方式，只有两种模式，设定在属性为 android:orientation 项目中。可以先将该属性项目的设定值清除后，按下"Alt"+"/"辅助输入。

　　• "horizontal"：由左至右地水平排列容器中的组件。

　　• "vertical"：由上至下，垂直排列容器中的组件。

　　• xmlns:android：表示以下的 XML 命名空间，通常设定在最顶端的组件即可，其他组件无须再做重复设定，除非有所不同。

　　• android:layout_width：各组件的版面配置宽度（大多数的组件都必须设定的项目）。

- fill_parent：填满该组件的父容器。
- match_parent：符合该组件的父容器（与 fill_parent 相同）。
- wrap_content：依照该组件内容决定。
- android:layout_height：各组件的版面配置高度(大多数的组件都必须设定的项目)。
- Button 是一种显示组件，就是显示一个按钮的组件，其实只是视觉上的按钮，实质上和 TextView 没有什么不同。在 Android 中的显示组件，全部都可以有按下的事件处理，不一定是按钮才有。
- TextView 用来显示一般文字数据内容。
- android:id：为该显示组件设定整个项目中的唯一识别名称，其格式为 "@+id/识别名称"，命名原则与 Java 变量相同，不可以是关键词或是保留字，而且 [a-zA-Z$_][a-zA-Z0-9$_]*，首字母可以是 a-zA-Z 或是 $、_ 字符，第二个字符之后可以多了数字使用。
- android:text：该组件的显示字符串内容。
- android:gravity：数据内容排列原则，常见如下。

center_horizontal：水平置中。

center_vertical：垂直置中。

center_vertical|center_horizontal：同时垂直水平置中。

修改过程中，随时按下"Ctrl"+"S"进行保存；按下"Ctrl+Shift"+"F"可以将文件以缩排架构排列。

查看目前的画面配置状况，点击"Graphical Layout"的页签，如下图所示。

## 2-1-3 开发程序

点击左边项目窗口中，"HelloLottery/"→"src/"→"tw.brad.android.book.hellolottery/"→"MainActivity.java"，中间出现以下已经写好的程序代码如下：

```
package tw.brad.android.book.hellolottery;

import android.os.Bundle;
import android.app.Activity;
import android.view.Menu;

public class MainActivity extends Activity {

    @Override
    protected void onCreate(Bundle savedInstanceState) {
        super.onCreate(savedInstanceState);
            setContentView(R.layout.activity_main);
    }

    @Override
    public boolean onCreateOptionsMenu(Menu menu) {
        // Inflate the menu; this adds items to the action bar if it is present.
        getMenuInflater( ).inflate(R.menu.main, menu);
        return true;
    }
}
```

还是这句话：不认识的不要放在自己的项目中。先修改成如下所示：

```
package tw.brad.android.book.hellolottery;

import android.os.Bundle;
import android.app.Activity;

public class MainActivity extends Activity {

    @Override
    protected void onCreate(Bundle savedInstanceState) {
            super.onCreate(savedInstanceState);
            setContentView(R.layout.activity_main);

    }
}
```

说明如下：

- package...：定义该Android程序所属的Package Name（通常不需要处理）。

- import…：定义该Android程序中所需要的API的Package Name（通常不需要处理，可以在开发中，按下"Ctrl+Shift+O"自动处理即可）。
- public class MainActivity extends Activity {…}：定义自定义的Android应用程序的类别名称，所有在应用程序中用户看到的部分，都是继承自Activity，所以会有extends Activity的定义；{…}程序区块则为该类别的详细定义。
- @Override：用来说明以下定义的方法是改写父类别的方法。
- protected void onCreate( ){}：用来开发撰写生命周期的第一个阶段应该要处理的程序内容。
- super.onCreate( )：用来调用父类别定义的生命周期的第一个阶段程序（必须）。
- setContentView(R.layiut.activity_main)：则是开始进行自行开发的内容中，不同的内容画面，指定前面所规划的版面内容。

### ■ 2-1-4 安装执行测试

迫不及待地执行看看吧，执行方式可以如下图所示，按下"Android Application"。

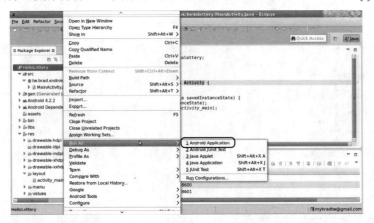

如果只连接一个实体装置或是一个已经启动完成的仿真器，则将会直接安装并执行到该装置上。

如果已经连接多个实体装置，或是一个实体装置及一个仿真器，则将会出现类似以下对话框，选择安装执行的特定装置。

点选仿真器之后，看到仿真器的状况如下。

至此，"Hello，World"算是完成了。

## 2-2 "BMI? Lottery！"

延续上一小节项目"Hello，Lottery"，将其完整地进行后续开发……

###  2-2-1 存取控制元件

项目中有两个元件Button和TextView。用户将会按下Button之后，出现一组乐透号码（然后买了一张，中了头奖，从此过着幸福快乐的生活……）。所以首要之务就是将两个元件的参考指针找出来。

先在MainActivity的类别中定义属性成员：

```
private Button torich;
private TextView richnum;
```

并在onCreate( )方法中，setContentView( )叙述句之后，以findView ById( )方法调用，传回其对象参考。从逻辑性来看，因为先有setContent View( )，整个版面内容设定之后，再从其中找出显示元件（所有显示元件都是View，也就是is-a View）。如果颠倒顺序，从java语言文法而言没有错误，却将会在运行时间抛出Exception而中断执行，这就是出现逻辑错误所造成的。

```
@Override
protected void onCreate(Bundle savedInstanceState) {
    super.onCreate(savedInstanceState);
    setContentView(R.layout.activity_main);

    torich = (Button)findViewById(R.id.torich);
    richnum = (TextView)findViewById(R.id.richnum);
}
```

读者在输入 R.id. 之后，稍候一下，应该会出现菜单，可以点选出当时在版面配置上所赋予的 android:id 的设定值，如果没有出现的话，有一种可能性就是尚未在版面配置文件编辑中存盘，以至于项目架构中尚未配置出该组件的对象参考指针资源。养成随时编辑，随时保存文件内容"Ctrl+S"的习惯。

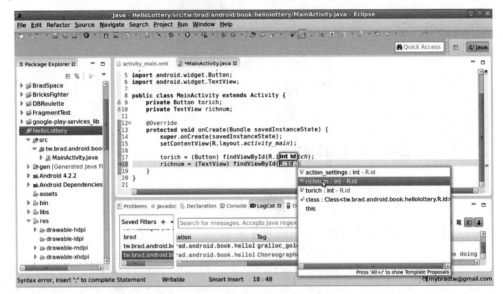

### ■ 2-2-2 按钮事件处理模式

按下 Button 之后的处理程序，就是为该 Button 组件设定按下事件的监听对象 (XxxListener)，而由该监听对象来处理按下之后该做的事。

```
torich.setOnClickListener( );
```

将会暂时出现底下红色波浪底线，表示语法错误，因为尚未传递所需要的监听对象参数。

增加其参数内容如下：

```
torich.setOnClickListener(new OnClickListener( ) {
    @Override
    public void onClick(View v) {
    // TODO Auto-generated method stub

    }
});
```

仍有错误来自于尚未 import，按下"Ctrl+Shift+O"，自动 import 处理。

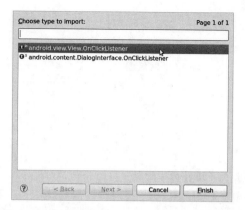

目前的监听对象是 View 的 OnClickListener，而不是 DialogInterface.OnClickListener，别选错了。观念就是 Button 是 View 组件，当然是以 View 的监听对象来负责监听触发事件。后面单元还会为读者介绍对话框的监听对象，那个时候就是使用 DialogInterface.OnClickListener 为其监听对象。

而在其中的 onClick( ) 方法中，开发撰写当用户按下按钮之后的处理程序。首先，先另外定义出处理的自定义方法，笔者的开发模式不喜欢全部塞在一起，容易造成维护不易的困扰，所以另外定义出自定义方法来处理，如下：

```java
package tw.brad.android.book.hellolottery;

import android.app.Activity;
import android.os.Bundle;
import android.view.View;
import android.view.View.OnClickListener;
import android.widget.Button;
import android.widget.TextView;

public class MainActivity extends Activity {
    private Button torich;
    private TextView richnum;

    @Override
    protected void onCreate(Bundle savedInstanceState) {
        super.onCreate(savedInstanceState);
        setContentView(R.layout.activity_main);

        torich = (Button) findViewById(R.id.torich);
        richnum = (TextView) findViewById(R.id.richnum);

        torich.setOnClickListener(new OnClickListener( ) {
```

```
                @Override
                public void onClick(View v) {
                        // 此处调用使用 createLottery( )方法
                        createLottery( );
                }
        });
}

// 该方法用来产生乐透号码
private void createLottery( ){

}
```

### 2-2-3 开发设计功能

好好专心来写createLottery( )方法吧……

```
// 该方法用来产生乐透号码
private void createLottery( ){
    HashSet<Integer> set = new HashSet<Integer>( );
    while (set.size( )<6){
        set.add((int)(Math.random( )*49+1));
    }
    richnum.setText("");
    Iterator<Integer> iterator = set.iterator( );
    while (iterator.hasNext( )){
        int num = iterator.next( );
        richnum.append(num + "  ");
    }
}
```

说明如下：

• HashSet用来声明一个可以存放数据的数据结构，利用数据不重复的特性（乐透号码也不会重复出现），并且泛型Integer。

• while( )循环侦测数据要是小于6个号码，继续执行到有6个不重复的号码。

• Math.random( )会传回大于或等于0.0与小于1.0之间的随机数，将该数乘以49加1后，其范围为大于或等于1.0到小于50.0（最大就是49.999……）之间的浮点数，再经过强制转型为int后就是介于1～49之间的随机数了。

- 将TextView对象变量richnum调用setText("")，将显示内容清除处理。
- HastSet对象实体set调用iterator( )方法传回Iterator对象实体，用来将数据内容依序取出使用。
- Iterator的hasNext( )方法传回boolean值表示是否还有数据存在。
- Iterator的hasNext( )方法传回数据内容。
- 最后由TextView的对象实体richnum调用append( )方法将数据一个一个连接出来。

执行看看吧……

当然，因为是随机数随机产生的结果，所以千万不要因为执行结果与此图不一样，就开始写邮件brad@brad.tw询问笔者。如果因此而中了大奖，那欢迎来信告知将会分给笔者一半奖金的消息，我应该不会拒绝您的善意。

### ■ 2-2-4　修改程序

开始有用户的抱怨声音出现：出现的号码没有排序，拿去彩票站不好兑奖……好，要用力倾听用户的使用经验，并进行改善。如果还要翻出当年差点被当掉的数据结构圣经来看的话，那可能下次改版时间将会是明年。此时善加运用Java中的TreeSet吧。

```java
// 该方法用来产生乐透号码
private void createLottery( ) {
    TreeSet<Integer> set = new TreeSet<Integer>( );
    while (set.size( ) < 6) {
        set.add((int) (Math.random( ) * 49 + 1));
    }
    richnum.setText("");
    Iterator<Integer> iterator = set.iterator( );
    while (iterator.hasNext( )) {
        int num = iterator.next( );
        richnum.append(num + "  ");
    }
}
```

只有将 HashSet 改成 TreeSet 而已，看看结果吧……

就不再重复说明了，我知道善良真诚的读者，您会知道知恩图报的。

## 2-3 写完了，然后呢？

写完了，然后呢？包装整理成为完整的 App 吧！

### 2-3-1 加上欢迎界面

"别人正式发布的 App，好像都有一个感觉还不错的欢迎界面，或是来一个开发公司的标记……"

那就来吧！先请专业的美工，不是美工，应该是视觉创意达人，设计了胜利标章。

利用准备好的图文件 winner.png 来设计欢迎界面，先在项目目录下 res/ 目录下建立出一个子目录，名称为"drawable"（全小写，请先按照我的命名 drawable）。并将图文件放在该目录下。

第 2 课　基本程序运行原理与应用　29

接着在 Package Name 下，右键单击点出开启建立 Activity 的对话框。

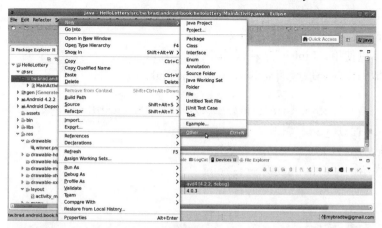

以下能直接按下"Next"按钮就按下吧。

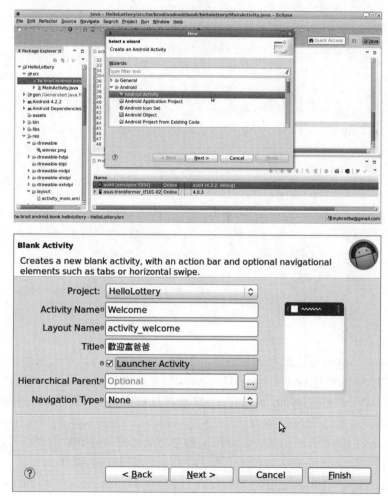

- Activity Name：输入欢迎界面的名称。
- Layout Name：会自动产生，当然也可以变更。
- Title：是显示在界面上方的文字内容。

看见可以使用"Finish"按钮，那就按下去！

一样地先设计处理版面 activity_welcome.xml。

```xml
<RelativeLayout xmlns:android="http://schemas.android.com/apk/res/android"
    android:layout_width="match_parent"
    android:layout_height="match_parent"
    android:background="#ffff00"
    >
    <TextView
        android:layout_width="wrap_content"
        android:layout_height="wrap_content"
        android:layout_centerHorizontal="true"
        android:layout_alignParentTop="true"
        android:text="亿万富翁大乐透"
        android:textColor="#0000ff"
        android:textSize="36sp"
        android:textStyle="bold|italic" />
    <ImageView
        android:id="@+id/weinner"
        android:layout_width="wrap_content"
        android:layout_height="wrap_content"
        android:layout_centerInParent="true"
        android:layout_centerHorizontal="true"
        android:src="@drawable/winner" />
</RelativeLayout>
```

说明如下：

- RelativeLayout 这次采用相对的版面配置，配置组件方式都是采用相对关系。
- 在 TextView 中的 layout_centerHorizontal 就是相对于父容器组件的水平置中。
- 在 TextView 中的 layout_allignParentTop 就是说明其配置的位置是沿着父容器组件的顶端。
- RelativeLayout 中的 background 用来指定背景处理，#ffff00 则是以 HTML 的颜色设定处理 RGB，目前显示红色加上绿色产生黄色作为背景颜色。
- ImageView 中的 src 则是用来指定呈现刚刚放在 res/drawable/ 下的 winner.png

的图文件。

## 2-3-2 调整启动程序

修改目前的项目目录下的 AndroidManifest.xml。

```xml
<?xml version="1.0" encoding="utf-8"?>
<manifest xmlns:android="http://schemas.android.com/apk/res/android"
    package="tw.brad.android.book.hellolottery"
    android:versionCode="1"
    android:versionName="1.0" >

    <uses-sdk
        android:minSdkVersion="8"
        android:targetSdkVersion="17" />

    <application
        android:allowBackup="true"
        android:icon="@drawable/ic_launcher"
        android:label="@string/app_name"
        android:theme="@style/AppTheme" >
        <activity
        android:name="tw.brad.android.book.hellolottery.MainActivity"
            android:label="@string/app_name" >
            <intent-filter>
            <action android:name="android.intent.action.MAIN" />

            <category android:name="android.intent.category.LAUNCHER" />
            </intent-filter>
        </activity>
        <activity
            android:name="tw.brad.android.book.hellolottery.Welcome"
            android:label="@string/title_activity_welcome" >
            <intent-filter>
             <action android:name="android.intent.action.MAIN" />
          <category android:name="android.intent.category.LAUNCHER" />
            </intent-filter>
        </activity>
    </application>
</manifest>
```

因为一开始只有 MainActivity 为启动项目的类别，因此后来加上的 Welcome 要成为启动应用程序的类别，可以看到各自文件架构下，都有以下内容：

```xml
<intent-filter>
    <action android:name="android.intent.action.MAIN" />
    <category android:name="android.intent.category.LAUNCHER" />
</intent-filter>
```

去掉 MainActivity 中的这个部分，整个改成以下结构：

```xml
<?xml version="1.0" encoding="utf-8"?>
<manifest xmlns:android="http://schemas.android.com/apk/res/android"
    package="tw.brad.android.book.hellolottery"
    android:versionCode="1"
    android:versionName="1.0" >

    <uses-sdk
        android:minSdkVersion="8"
        android:targetSdkVersion="17" />

    <application
        android:allowBackup="true"
        android:icon="@drawable/ic_launcher"
        android:label="@string/app_name"
        android:theme="@style/AppTheme" >
        <activity
          android:name="tw.brad.android.book.hellolottery.MainActivity"
          android:label="@string/app_name" >
        </activity>
        <activity
            android:name="tw.brad.android.book.hellolottery.Welcome"
            android:label="@string/title_activity_welcome" >
            <intent-filter>
            <action android:name="android.intent.action.MAIN" />
                <category android:name="android.intent.category.LAUNCHER" />
            </intent-filter>
        </activity>
    </application>
</manifest>
```

开发撰写 Welcome.java 如下所示：

```java
package tw.brad.android.book.hellolottery;
import android.app.Activity;
import android.content.Intent;
import android.os.Bundle;
import android.view.View;
import android.view.View.OnClickListener;
import android.widget.ImageView;
public class Welcome extends Activity {
    private ImageView winner;

    @Override
    protected void onCreate(Bundle savedInstanceState) {
        super.onCreate(savedInstanceState);
        setContentView(R.layout.activity_welcome);

        winner = (ImageView)findViewById(R.id.winner);
        winner.setOnClickListener(new OnClickListener( ) {
            @Override
            public void onClick(View v) {
                gotoMain( );
            }
        });
    }

    private void gotoMain( ){
        Intent intent = new Intent(this, MainActivity.class);
        startActivity(intent);
        finish( );
    }
}
```

处理结构与产生乐透非常类似，以下说明点按胜利图片之后的部分：

• Intent 类别对象用来设定调用其他 Activity 或是 Service 的对象实体。

• 调用 startActivity( )，并传入设定好的 Intent 对象实体，即可将控制权转移到 MainActivity 类别对象了。

• 之后不再需要看到欢迎画面，于是调用 finish( )。

结果如下：

## 2-4 Activity的生命周期

一个简单的项目开发完成后，开始对Android的App开发详细部分进行了解。

事实上，对于开发过网页设计并搭配JavaScript处理动态效果的读者来说，应该感受到两者异曲同工之处了。处理版面配置就如同设定规划HTML的网页内容，将要程序控制的元素赋予id；而开发程序部分就如同处理JavaScript的部分。甚至还调用findViewById( )的方法，几乎和JavaScript中的getElementById( )没有两样，算是非常容易学习的。

### ■ 2-4-1 生命周期的观念

生命周期只是将Activity从启动后，执行、暂停、恢复到结束的过程，看成为一个生命周期。

在Android App中的Activity扮演的角色，其实已经在前几小节中看到作用了。只要是用户看得到的部分，都是由Activity负责进行控制处理（如同JavaScript负责的部分）。

在google的开发者官方网站上有一个非常重要的图，呈现出一个Activity的生命周期：

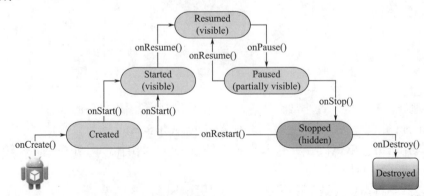

一般启动Activity的基本程序如下：

- onCreate( )
- onStart( )
- onResume( )

之后就进入到执行状态。当用户按下返回键之后，会使该Activity从执行状态中执行以下程序：

- onPause( )
- onStop( )
- onDestroy( )

最后进入到摧毁完毕阶段。如果在执行状态下，启动其他的Activity之后，会使得目前的Activity执行以下程序：

- onPause( )
- onStop( )

而进入到暂停状态，当再度被恢复执行时，将会执行以下程序：

- onRestare( )
- onStart( )
- onResume( )

再度回到执行状态中。

从Activity的API中可以看到另外一张示意图。

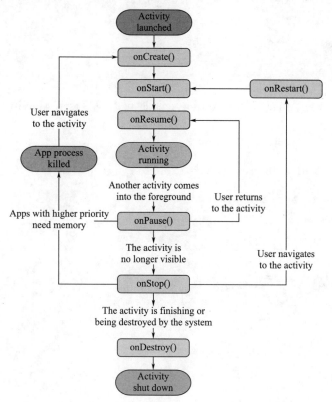

## 2-4-2 测试实作

可以开发一个简单的测试项目来实作生命周期：
处理版面配置res/layout/activity_main.xml。

```xml
<LinearLayout xmlns:android="http://schemas.android.com/apk/res/android"
    xmlns:tools="http://schemas.android.com/tools"
    android:layout_width="match_parent"
    android:layout_height="match_parent"
    android:orientation="vertical"
    >

    <Button
        android:id="@+id/next_page"
        android:layout_width="match_parent"
        android:layout_height="wrap_content"
        android:text="Goto Next Page" />

    <TextView
        android:layout_width="wrap_content"
        android:layout_height="wrap_content"
        android:text="Main Page" />

</LinearLayout>
```

按钮设计是为了启动另一个Activity，此时可以观察目前Activity的运行状况。

src/MainActivity.java

```java
package tw.brad.android.book.helloactivity;

import android.app.Activity;
import android.content.Intent;
import android.os.Bundle;
import android.util.Log;
import android.view.View;
import android.view.View.OnClickListener;
import android.widget.Button;

public class MainActivity extends Activity {
    private Button next_page;
```

```java
@Override
protected void onCreate(Bundle savedInstanceState) {
    super.onCreate(savedInstanceState);
    setContentView(R.layout.activity_main);
    Log.i("brad", "onCreate");

    next_page = (Button)findViewById(R.id.next_page);
    next_page.setOnClickListener(new OnClickListener( ) {
        @Override
        public void onClick(View v) {
            Intent itent = new Intent(MainActivity.this,
            Page2Activity.class);
            startActivity(itent);
        }
    });
}

@Override
protected void onDestroy( ) {
    super.onDestroy( );
    Log.i("brad", "onDestroy");
}

@Override
protected void onPause( ) {
    super.onPause( );
    Log.i("brad", "onPause");
}

@Override
protected void onRestart( ) {
    super.onRestart( );
    Log.i("brad", "onRestart");
}

@Override
protected void onResume( ) {
    super.onResume( );
    Log.i("brad", "onResume");
}
```

```
    @Override
    protected void onStart( ) {
        super.onStart( );
        Log.i("brad", "onStart");
    }
    @Override
    protected void onStop( ) {
        super.onStop( );
        Log.i("brad", "onStop");
    }
}
```

说明如下：

• 在各个onXxx( )方法中调用Log.i("brad"，"Xxx")以调用产生LogCat的观察显示。

• 按下按钮启动下一个Activity，观看目前Activity的运行状况。

LogCat视图如果没有显示，可以在Eclipse的菜单列点选Window后下拉菜单中选择"Show View...Other"。

应该就会显示出LogCat的页签，点击之后，可以在左侧中的加号点击后，输入要过滤的Log数据。

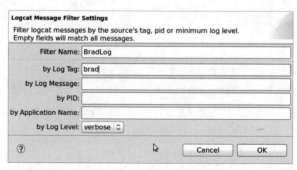

这样就会过滤出 App 执行中，Log 调用传入字符串为 "brad" 的 tag 信息数据。

### 2-4-3 开始观察

执行启动之后，看到右边的 LogCat 记录内容。

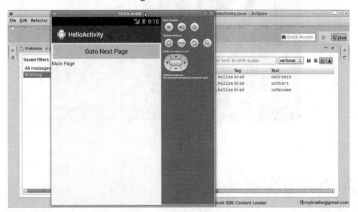

按下按钮来启动下一个 Activity。

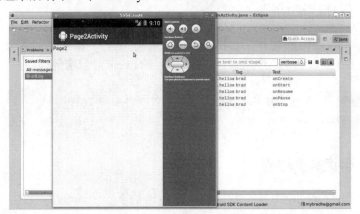

再从 Page2Activity 按下返回键回到 MainActivity。

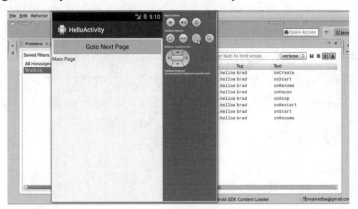

最后结束 MainActivity。

 **2-5　Activity切换Activity**

### ■ 2-5-1　仅作启动切换

不同 Activity 之间的切换，已经在前几小节中认识。就是通过设定 Intent 对象实体，并调用 startActivity( ) 方法启动传入参数的 Intent 对象实体即可。如下：

```
Intent itent = new Intent(MainActivity.this,Page2Activity.class);
startActivity(itent);
```

### ■ 2-5-2　传递数据过去

当从原来的 Activity 中要将特定的数据传递给下一个 Activity 处理或是判断时，则可以通过 Intent 对象实体进行设定。调用 putExtra("Key"，Xxx 形式数据值) 即可，例如：

```
private void gotoPage2( ){
    Intent intent = new Intent(this, Page2Activity.class);
    intent.putExtra("出版", "计算机人");
    intent.putExtra("作者", "赵令文");
    intent.putExtra("isAndroid", true);
    intent.putExtra("price", 10000);

    startActivity(intent);
}
```

而在Page2Activity中，先调用Activity的getIntent( )方法传回由之前Activity所传入的Intent对象实体。再由该对象实体调用getXxxExtra( )方法，传入指定字符串"Key"，而将其数据值传回，如果不存在该指定字符串"Key"的数据，则第二个参数设定其默认值。

```java
package tw.brad.android.book.a2a;

import android.app.Activity;
import android.content.Intent;
import android.os.Bundle;
import android.widget.TextView;

public class Page2Activity extends Activity {
    private TextView page2_msg;

    @Override
    protected void onCreate(Bundle savedInstanceState) {
        super.onCreate(savedInstanceState);
        setContentView(R.layout.activity_page2);

        page2_msg = (TextView)findViewById(R.id.page2_msg);
        Intent it = getIntent( );
        String pub = it.getStringExtra("出版");
        String auth = it.getStringExtra("作者");
        int price = it.getIntExtra("price", 0);
        boolean isAndroid = it.getBooleanExtra("isAndroid", false);
        page2_msg.setText("出版社: " + pub + "\n" +
                    "作者: " + auth + "\n" +
                    "售价:" + price + "\n" +
                    "Android:" + (isAndroid?"Yes":"No"));
    }
}
```

### ■ 2-5-3　切换之后回来确认

如果只是单纯的一对一，则不用判断就知道是哪个Activity切换回来。而如果是一对多的切换，而回来之后需要知道分别是哪个Activity切换回来，并继续针对差异处理，那么就必须在激活切换的时候，改调用startActivityForResult( )，并设定一个自订的要求码（Request Code）。

```
private void gotoPage2( ) {
    Intent intent = new Intent(this, Page2Activity.class);
    intent.putExtra("出版", "计算机人");
    intent.putExtra("作者", "赵令文");
    intent.putExtra("isAndroid", true);
    intent.putExtra("price", 10000);
//      startActivity(intent);
    startActivityForResult(intent, 2);
}
private void gotoPage3( ) {
    Intent intent = new Intent(this, Page3Activity.class);
//      startActivity(intent);
    startActivityForResult(intent, 3);
}
```

并且同时也改写 Activity 的 onActivityResult( ) 方法。当由其他 Activity 返回此 Activity 时会执行该方法，传递当时过去的 RequestCode 回来，因此就可以判断如何区分之后的后续工作。例如照完相片回来之后观看影像文件，与录音完毕之后要聆听录音文件等，是不同的处理模式。

```
@Override
protected void onActivityResult(int requestCode, int resultCode, Intent data) {
    super.onActivityResult(requestCode, resultCode, data);
    if (requestCode == 2){
        msg.setText("Come back from Page2");
    }else if (requestCode == 3){
        msg.setText("Come back from Page3");
    }else {
        msg.setText("Hello, Page");
    }
}
```

### ■ 2-5-4　将数据传递回来

再来就是如何将激活切换回原来的 Activity，也顺便将新数据带回来。处理重点就是在切换过去的 Activity 在结束之前，调用 setResult( ) 方法，通过 Intent 对象实体将数据带回来，原本的 Activity 也是改写 onActivityResult( ) 方法接收参数处理。

可以只传递自定义结果码 Result Code(int 形式)：

```
@Override
public void finish( ) {
    setResult(RESULT_OK);
    super.finish( );
}
```

或是顺便通过 Intent 对象实体带数据：

```
@Override
public void finish( ) {
Intent it = new Intent( );
    it.putExtra("data", "OK");
    setResult(RESULT_OK, it);
    super.finish( );
}
```

而回到原来的 Activity 中：

```
@Override
protected void onActivityResult(int requestCode, int resultCode, Intent data) {
    super.onActivityResult(requestCode, resultCode, data);
    if (requestCode == 2) {
        msg.setText("Come back from Page2");
        if (resultCode == RESULT_OK){
            String result = data.getStringExtra("data");
        msg.append("\n" + result);
        }
    } else if (requestCode == 3) {
        msg.setText("Come back from Page3");
    } else {
        msg.setText("Hello, Page");
    }
}
```

一般应用程序大多数是一个以上的 Activity 在运行，因此本小节的内容算是非常基本的。

## 2-6 Service的运行应用

Activity负责处理用户接口的部分，而Service则是负责背景中执行的程序，通常很常见的方式是应用在有周期性的工作任务，例如游戏中的背景音乐播放，或是每隔一段时间就向远程服务器发出询问要求，当有特定的数据从远程服务器传回移动装置之后，可能会发出消息通知。类似这样的工作程序内容，不需要有任何用户接口的处理及运行，因此就可以以Service的模式进行开发。

因为上述的Service的背景运行执行特性，Service完全无法直接存取操作用户接口，可以通过android.os.Handler类别对象实体来进行存取前景的显示内容，或是以Broadcast的模式发出广播数据给Activity处理。

Service的生命周期通常会以Activity来进行启动或是绑定，也可以在Activity的执行期间结束Service的生命周期，但是并不会在Activity结束生命周期的同时也结束Service的生命周期。所以，经常看到在一般移动装置中的应用程序，看似已经按下返回键结束特定的应用程序，事实上该应用程序可能早就在Activity的执行期间已经激活Service的运行。所以用户可能已经结束执行程序，但是等一下刚才明明已经结束离开的应用程序可能会跳出一个通知，告诉你一个好消息："Brad要出新专辑了"。

Service不是分开的程序，也不属于线程，但是可以和线程共同合作演出。

### ■ 2-6-1 生命周期实测

先直接以一个Activity中通过两个Button来负责启动Service，及结束Service进行实测。

```
res/layout/activity_main.xml
<LinearLayout xmlns:android="http://schemas.android.com/apk/res/android"
    android:layout_width="match_parent"
    android:layout_height="match_parent"
     android:orientation="vertical"
     >

    <Button
        android:id="@+id/start"
        android:layout_width="match_parent"
        android:layout_height="wrap_content"
        android:text="Start Service"
        />
```

```xml
    <Button
        android:id="@+id/stop"
        android:layout_width="match_parent"
        android:layout_height="wrap_content"
        android:text="Stop Service"
        />
    <TextView
        android:id="@+id/msg"
        android:layout_width="wrap_content"
        android:layout_height="wrap_content"
        />
</LinearLayout>
```

而在 MainActivity.java 中：

```java
package tw.brad.android.book.myservicetest;
import android.app.Activity;
import android.content.Intent;
import android.os.Bundle;
import android.view.View;
import android.view.View.OnClickListener;
import android.widget.Button;
import android.widget.TextView;
public class MainActivity extends Activity {
    private Button start, stop;
    private TextView msg;

    @Override
    protected void onCreate(Bundle savedInstanceState) {
    super.onCreate(savedInstanceState);
        setContentView(R.layout.activity_main);

        msg = (TextView)findViewById(R.id.msg);

        start = (Button)findViewById(R.id.start);
        start.setOnClickListener(new OnClickListener( ) {
            @Override
```

```
            public void onClick(View v) {
                startMyService( );
            }
        });
        stop = (Button)findViewById(R.id.stop);
        stop.setOnClickListener(new OnClickListener( ) {
            @Override
            public void onClick(View v) {
                stopMyService( );
            }
        });
    }
    // 用来负责启动 MyService
    private void startMyService( ){
        Intent it= new Intent(this, MyService.class);
        startService(it);
    }
    // 用来负责结束 MyService
    private void stopMyService( ){
        Intent it= new Intent(this, MyService.class);
        stopService(it);
    }
}
```

启动Service的方式非常简单，与启动另一个Activity的模式非常类似，还是通过Intent的对象实体，设定要启动的Service类别，接着调用startService( )方法，而将设定好的Intent对象实体带入参数即可。

以下是测试生命周期的Service：

MyService.java

```
package tw.brad.android.book.myservicetest;

import android.app.Service;
import android.content.Intent;
import android.os.IBinder;
import android.util.Log;
```

```
public class MyService extends Service {
    public MyService( ) {
    }

    @Override
    public IBinder onBind(Intent intent) {
        // TODO: Return the communication channel to the service.
        throw new UnsupportedOperationException("Not yet implemented");
    }

    @Override
    public void onCreate( ) {
        super.onCreate( );
        Log.i("brad", "onCreate");
    }

    @Override
    public int onStartCommand(Intent intent, int flags, int startId) {
        Log.i("brad", "onStartCommand");
        return super.onStartCommand(intent, flags, startId);
    }

    @Override
    public void onDestroy( ) {
        super.onDestroy( );
        Log.i("brad", "onDestroy");
    }
}
```

开始进行测试了解……

第一次按下 Start：

- onCreate( )
- onStartCommand( )

再度按下 Start：

- onStartCommand( )

接着按下 Stop：

- onDestroy( )

当第一次启动 Service 时，会先执行 onCreate( )，之后重复调用执行 startService( ) 时，则只会再度执行 onStartCommand( )，而不再执行 onCreate( )，表示可以通过 startService( ) 的调用传递不同的 Intent 数据内容给 Service 进行处理。而当已经 stopService( ) 之后，Service 会执行 onDestroy( )，再度 startService( )，则会从 onCreate( ) 开始执行。

## ■ 2-6-2　与线程共舞

接着在 MyService 中启动一个简单的 java.util.TimerTask，以周期性执行特定的工作任务。

```java
package tw.brad.android.book.myservicetest;
import java.util.Timer;
import java.util.TimerTask;

import android.app.Service;
import android.content.Intent;
import android.os.IBinder;
import android.util.Log;
public class MyService extends Service {
    private Timer timer;
    private MyTask task;
    private int i;
    public MyService( ) {
        timer = new Timer( );
    }

    @Override
    public IBinder onBind(Intent intent) {
        // TODO: Return the communication channel to the service.
        throw new UnsupportedOperationException("Not yet implemented");
    }

    private class MyTask extends TimerTask{
        @Override
        public void run( ) {
            i++;
        }
    }

    @Override
    public void onCreate( ) {
```

```
        super.onCreate( );
        task = new MyTask( );
        timer.schedule(task, 100, 400);
    }

    @Override
    public int onStartCommand(Intent intent, int flags, int startId) {
        i = intent.getIntExtra("i", -1);
        return super.onStartCommand(intent, flags, startId);
    }

    @Override
    public void onDestroy( ) {
        super.onDestroy( );
        timer.cancel( );
    }
}
```

当按下Start后，后台中启动了Service，并且开始Timer的周期任务执行MyTask对象实体，表示已经开始运行，而当再度重复按下Start时，则会变更目前Service对象实体的属性的i值，周期运作状况照旧，直到按下Stop后取消Timer的所有周期任务的执行。

目前看到的是前台操控后台的Service。

### 2-6-3　通过Broadcast发送数据给前台

先在Service中的适当时机，将数据放在一个Intent对象实体，再调用sendBroadcast( )方法将该Intent对象实体广播出去。而intent对象实体会设定一个字符串Action（范例中的"fromMyService"）。等一下在接收Broadcast时会用此Action字符串进行Intent的过滤机制。

```
MySerview.java
package tw.brad.android.book.myservicetest;

import java.util.Timer;
import java.util.TimerTask;

import android.app.Service;
import android.content.Intent;
import android.os.IBinder;
```

```java
import android.util.Log;
public class MyService extends Service {
    private Timer timer;
    private MyTask task;
    private int i;
    private Intent it2a;
    public MyService( ) {
        timer = new Timer( );
        it2a = new Intent("fromMyService");
    }

    @Override
    public IBinder onBind(Intent intent) {
       // TODO: Return the communication channel to the service.
        throw new UnsupportedOperationException("Not yet implemented");
    }

    private class MyTask extends TimerTask{
        @Override
        public void run( ) {
            i++;
            it2a.putExtra("i", i);
            sendBroadcast(it2a);
        }
    }

    @Override
    public void onCreate( ) {
        super.onCreate( );
        task = new MyTask( );
        timer.schedule(task, 100, 400);
    }

    @Override
    public int onStartCommand(Intent intent, int flags, int startId) {
        i = intent.getIntExtra("i", -1);
        return super.onStartCommand(intent, flags, startId);
    }
```

```
    @Override
    public void onDestroy( ) {
        super.onDestroy( );
        timer.cancel( );
    }
}
```

回到前台的Activity处理如下所示。

• 开发一个自订类别继承android.content.BroadcastReceiver。

• 改写onReceive( )方法，针对传进来的Intent对象实体参数，建构出该Broadcast-Receiver对象实体。

• 建构一个IntentFilter对象实体，设定Action字符串，与service端配合使用

• 由Activity调用registerReceiver( )方法，传入BroadcastReceiver对象实体与intentFilter对象实体。

MainActivity.java

```java
package tw.brad.android.book.myservicetest;
import android.app.Activity;
import android.content.BroadcastReceiver;
import android.content.Context;
import android.content.Intent;
import android.content.IntentFilter;
import android.os.Bundle;
import android.view.View;
import android.view.View.OnClickListener;
import android.widget.Button;
import android.widget.TextView;
public class MainActivity extends Activity {
    private Button start, stop;
    private TextView msg;
    private MyReceiver mr;

    @Override
    protected void onCreate(Bundle savedInstanceState) {
        super.onCreate(savedInstanceState);
        setContentView(R.layout.activity_main);

        msg = (TextView)findViewById(R.id.msg);
```

```java
            start = (Button)findViewById(R.id.start);
            start.setOnClickListener(new OnClickListener( ) {
                @Override
                public void onClick(View v) {
                    startMyService( );
                }
            });
            
            stop = (Button)findViewById(R.id.stop);
            stop.setOnClickListener(new OnClickListener( ) {
                @Override
                public void onClick(View v) {
                    stopMyService( );
                }
            });
            
            mr = new MyReceiver( );
            registerReceiver(mr, new IntentFilter("fromMyService"));
    }
    
    // 用来负责启动 MyService
    private void startMyService( ){
        Intent it= new Intent(this, MyService.class);
        it.putExtra("i", (int)(Math.random( )*49+1));
        startService(it);
    }
    
    // 用来负责结束 MyService
    private void stopMyService( ){
        Intent it= new Intent(this, MyService.class);
        stopService(it);
    }
    
    private class MyReceiver extends BroadcastReceiver {
        @Override
        public void onReceive(Context context, Intent intent) {
            int i = intent.getIntExtra("i", -1);
            msg.setText("i = " + i);
        }
    }
}
```

# 03 Chapter

# 第3课 基本用户界面与事件触发

3-1 条列显示元件 ListView

3-2 线性配置 LinearLayout

3-3 相对配置 RelativeLayout

3-4 表格配置 TableLayout

3-5 网格显示 GridView

3-6 滑动显示 ViewFlipper

## 3-1 条列显示元件ListView

大部分的Android初学书籍或是网络文章，都不太会将ListView放在一开始介绍，因为会有些许难度。笔者认为反正早晚都会面对，又不是一定要先学会较简单的LinearLayout或是RelativeLayout才能够了解ListView，再加上ListView可用来当作第一个画面，读者可以运用第2课学习到的Activity之间的切换，练习切换到其他的版面配置，往后这个项目App可以留着参考或是增加更多学习的版面配置，如此一来更具有学习上的实用价值意义，因此，就开始吧！

### ■ 3-1-1 基本格式

这次并非先从版面配置下手，而是直接回到Activity设计下手。先将extends Activity改成ListActivity。其实ListActivity从名称来看大约可以判断其为Activity的子类别，所以具有与Activity相同的生命周期，特色就是本身直接含有一个ListView组件。直接调用getListView( )将会回传本身的ListView对象实体。并且记得将setContentView( )那一列删除掉，因为整个画面就是一个ListView，不再需要另外指定。

MainActivity.java

```java
package tw.brad.book.hellolayout;
import android.app.ListActivity;
import android.os.Bundle;
import android.widget.ListView;
public class MainActivity extends ListActivity {
    private ListView lview;

    @Override
    protected void onCreate(Bundle savedInstanceState) {
        super.onCreate(savedInstanceState);
        lview = getListView( );
    }
}
```

接着下面的处理重点在于条列的单一项目版面配置（不是整个画面）。就先用原来已经存在的activity_main.xml，如果嫌名称不好，可在其上按下鼠标右键"Refactor"→"Rename"更改。笔者换成"menuitem.xml"。

内容略作些许修改，因为要当成用户点按的项目，所以将字体略作调整。

menuitem.xml：

```xml
<RelativeLayout xmlns:android="http://schemas.android.com/apk/res/android"
    android:layout_width="match_parent"
    android:layout_height="wrap_content"
    android:paddingBottom="@dimen/activity_vertical_margin"
    android:paddingLeft="@dimen/activity_horizontal_margin"
    android:paddingRight="@dimen/activity_horizontal_margin"
    android:paddingTop="@dimen/activity_vertical_margin">

    <TextView
        android:id="@+id/item_txt"
        android:layout_width="match_parent"
        android:layout_height="wrap_content"
        android:textSize="24sp" />

</RelativeLayout>
```

再度回到 MainActivity.java 中。

ListView 可以通过调用 setAdapter( ) 方法将数据以条列方式呈现，呈现版面如上所示的 menuitem.xml，接下来就是处理这中间的关系。负责串起中间关系的关键就是实作 android.widget.ListAdapter 的对象实体，可以经由 android.widget.SimpleAdapter 来建构出来。要显示的数据以字符串数组存放，全部数据放在 java.util.List 的数据结构中，每个元素会通过 java.util.Map 的数据的 Key 将值对应到版面中 id 为 item_txt 的组件内容上。

声明定义：

```java
private SimpleAdapter adapter;
private String[] items = {
    "LinearLayout", "RelativeLayout", "TableLayout"
    };
private String[] from = {"item_txt"};
private int[] to = {R.id.item_txt};
private ArrayList<HashMap<String,String>> data;
```

再在 onCreate( ) 中：

① 建构出 data 的对象实体。

② 设计 for-each 循环将 items[ ] 的数据逐笔跑过。
③ 每跑一笔 item，先建构出单一的 HashMap 数据结构对象实体。
④ 放进 Key 为 from[0] 的字符串，其值为 item。
⑤ 再将 HashMap 对象加进 data 中。
⑥ 跑完 for-each 循环后，就可以建构出 SimpleAdapter 对象实体了。
⑦ 最后 ListView 设定 adapter 即可。

```
@Override
protected void onCreate(Bundle savedInstanceState) {
    super.onCreate(savedInstanceState);

    lview = getListView( );

    data = new ArrayList<HashMap<String,String>>( );
    for (String item : items){
        HashMap<String,String> temp = new HashMap<String, String>( );
        temp.put(from[0], item);
        data.add(temp);
    }
    adapter = new SimpleAdapter(this, data, R.layout.menuitem, from, to);

    lview.setAdapter(adapter);
}
```

应该会有以下画面呈现出来：

应该是相当容易处理的用户界面。

接着处理重点在于按下单一选项的动作处理。处理观念就是针对 ListView 对象设定调用 setOnItemClickListener( ) 将选项按下，监听对象参数传入即可。

下面这段处理方式是将传入的第三个 int 参数，以显示 String[index] 方式取得，并通过 Toast 方式显示出来。

```
lview.setOnItemClickListener(new OnItemClickListener( ) {
    @Override
    public void onItemClick(AdapterView<?> aview, View view, int index,
        long arg3) {
        Toast.makeText(MainActivity.this, items[index], Toast.
            LENGTH_LONG).show( );
    }
});
```

当使用按下第二个选项 RelativeLayout 之后，显示如下结果：

## 3-1-2 进阶格式

写到这里读者一定会觉得单一选项的内容过于单调，应该可以弄得丰富一点。

假设会有个生动活泼的图标，有标题列含有简单说明列。也先将准备好的图标放在 res/drawable/ 子目录下，如果没有存在 drawable/ 子目录的话，直接新增一个吧（记得大小写严格区分）。

所以先回到版面配置单一选项中处理：

```
<RelativeLayout xmlns:android="http://schemas.android.com/apk/res/android"
    android:layout_width="match_parent"
    android:layout_height="wrap_content"
    android:paddingBottom="@dimen/activity_vertical_margin"
    android:paddingLeft="@dimen/activity_horizontal_margin"
    android:paddingRight="@dimen/activity_horizontal_margin"
    android:paddingTop="@dimen/activity_vertical_margin" >
```

```xml
    <ImageView
        android:id="@+id/item_img"
        android:layout_width="wrap_content"
        android:layout_height="wrap_content"
        android:layout_alignParentLeft="true"
        android:layout_alignParentTop="true" />
    <TextView
        android:id="@+id/item_txt"
        android:layout_width="match_parent"
        android:layout_height="wrap_content"
        android:layout_alignParentTop="true"
        android:layout_toRightOf="@id/item_img"
        android:textSize="24sp" />
    <TextView
        android:id="@+id/item_desc"
        android:layout_width="match_parent"
        android:layout_height="wrap_content"
        android:layout_below="@id/item_txt"
        android:layout_toRightOf="@id/item_img"
        android:textSize="16sp" />
</RelativeLayout>
```

增加一个ImageView(id=item_img)及一个TextView(id=item_desc)。

回到MainActivity.java中,因为增加了两个项目,所以也增加了两组数组来存放。

```java
private String[] items = { "LinearLayout", "RelativeLayout", "TableLayout" };
private String[] items_desc = { "线性配置:针对组件进行垂直式或是水平式的排列",
    "相对配置:其所属的所有组件都是依照相对位置摆设",
    "表格配置:配置组件的原理与表格相同" };
private int[] items_img = { R.drawable.item0, R.drawable.item1,
R.drawable.item2 };
```

同时,对应关系的from[]与int[]也增加对应项目元素:

```java
private String[] from = { "item_txt", "item_desc", "item_img" };
private int[] to = { R.id.item_txt, R.id.item_desc, R.id.item_img };
```

而原本声明泛型<String,String>，也要修改为<Stirng,Object>，因为不再只有String的数据，还包含了整数（Integer）形式的数据（Auto-boxing）。

```
private ArrayList<HashMap<String, Object>> data;
```

因此在整理数据的程序代码也进行了些许变动：

```
data = new ArrayList<HashMap<String, Object>>( );
for (int i=0; i < items.length; i++){
    HashMap<String, Object> temp = new HashMap<String, Object>( );
    temp.put(from[0], items[i]);
    temp.put(from[1], items_desc[i]);
    temp.put(from[2], items_img[i]);
    data.add(temp);
}
adapter = new SimpleAdapter(this, data, R.layout.menuitem, from, to);
```

执行起来就将会有以下的呈现效果。

## 3-2 线性配置LinearLayout

LinearLayout的线性配置版面方式，应该算是所有版面配置中最简单，也是最常用最好用的配置方式。只有两种方式配置，不是垂直就是水平，从上至下或是从左至右。虽然如此，却有一个非常好用的权重属性，可以以比例方式来配置所属组件所占有的空间。

先来看一个基本 LinearLayout 配置架构：

```xml
<?xml version="1.0" encoding="utf-8"?>
<LinearLayout xmlns:android="http://schemas.android.com/apk/res/android"
    android:layout_width="match_parent"
    android:layout_height="match_parent"
    android:orientation="vertical" >
```

• </LinearLayout>

xmlns:XXX=""→用来定义该组件的 XML tag 的命名空间，通常不需要特别去处理。XXX 是可以自订名称，自定义名称之后，后面的属性应该就是 xxx:属性名称，目前预设为 android 就可以。

• layout_width 与 layout_width 标示该组件配置的宽高，通常在最外层的 Layout 应该都是 match_parent，而这两个属性是大多数组件都必须设定的项目。

• orientation 表示配置方向，不是 vertical（垂直）就是 horizontal（水平）。

现在加上一个组件（都先以 TextView 作为内容组件）：

```xml
<?xml version="1.0" encoding="utf-8"?>
<LinearLayout xmlns:android="http://schemas.android.com/apk/res/android"
    android:layout_width="match_parent"
    android:layout_height="match_parent"
    android:orientation="vertical" >

    <TextView
        android:layout_width="wrap_content"
        android:layout_height="wrap_content"
        android:background="#00ff00"
        android:text="Hello, Android"
    />

</LinearLayout>
```

TextView 加上 background 属性不是为了好看，而是可以帮助学习了解元件配置的空间范围。

目前被放在左上角。

如果只有一个组件，其实也可以省略 LinearLayout 的 orientation 的属性；但是两个以上就不可以省略了。

再加上一个 TextView，并使其各自宽度占满 LinearLayout 这个父容器。

```xml
<?xml version="1.0" encoding="utf-8"?>
<LinearLayout xmlns:android="http://schemas.android.com/apk/res/android"
    android:layout_width="match_parent"
    android:layout_height="match_parent"
    android:orientation="vertical" >

    <TextView
        android:layout_width="match_parent"
        android:layout_height="wrap_content"
        android:background="#00ff00"
        android:text="Hello, Android1"
    />

    <TextView
        android:layout_width="match_parent"
        android:layout_height="wrap_content"
        android:background="#ffff00"
        android:text="Hello, Android2"
    />
</LinearLayout>
```

呈现出来的效果如下：

再来在各自的属性中加上 layout_weight，并设定其值为"1"。

```xml
<?xml version="1.0" encoding="utf-8"?>
<LinearLayout xmlns:android="http://schemas.android.com/apk/res/android"
    android:layout_width="match_parent"
```

```xml
    android:layout_height="match_parent"
    android:orientation="vertical" >
<TextView
    android:layout_width="match_parent"
    android:layout_height="wrap_content"
    android:background="#00ff00"
    android:text="Hello, Android1"
    android:layout_weight="1"
 />
<TextView
    android:layout_width="match_parent"
    android:layout_height="wrap_content"
    android:background="#ffff00"
    android:text="Hello, Android2"
    android:layout_weight="1"
 />
</LinearLayout>
```

呈现效果就是两个组件填满父容器，各自以1：1的比例占有空间。

如果想要配置上下组件各为1：4的比例，则设定如下：

```xml
<?xml version="1.0" encoding="utf-8"?>
<LinearLayout xmlns:android="http://schemas.android.com/apk/res/android"
    android:layout_width="match_parent"
    android:layout_height="match_parent"
```

```
        android:orientation="vertical" >
    <TextView
        android:layout_width="match_parent"
        android:layout_height="wrap_content"
        android:background="#00ff00"
        android:text="Hello, Android1"
        android:layout_weight="1"
        />
    <TextView
        android:layout_width="match_parent"
        android:layout_height="wrap_content"
        android:background="#ffff00"
        android:text="Hello, Android2"
        android:layout_weight="4"
        />
</LinearLayout>
```

呈现效果如下：

善加运用layout_weight的比例配置，可以适用于多种不同屏幕规格尺寸的移动装置上。

## 3-3 相对配置RelativeLayout

RelativeLayout的版面配置就是其所属的组件，配置的位置全部都是以相对的方式进行指派。可能会相对于父容器的上下左右，或是在特定已经完成指派位置之组件的上下左右。这样的配置方式也比较能够符合大多数的屏幕尺寸。

先来进行基本的架构：

```xml
<?xml version="1.0" encoding="utf-8"?>
<RelativeLayout xmlns:android="http://schemas.android.com/apk/res/android"
    android:layout_width="match_parent"
    android:layout_height="match_parent" >

</RelativeLayout>
```

直接先放两个背景颜色不一样的 TextView 组件上去。先是一个"Hello，World and Brad"，再来一个"Hello，Android"。

```xml
<?xml version="1.0" encoding="utf-8"?>
<RelativeLayout xmlns:android="http://schemas.android.com/apk/res/android"
    android:layout_width="match_parent"
    android:layout_height="match_parent" >

    <TextView
        android:layout_width="wrap_content"
        android:layout_height="wrap_content"
        android:background="#ffff00"
        android:text="Hello, World and Brad" />

    <TextView
        android:layout_width="wrap_content"
        android:layout_height="wrap_content"
        android:background="#00ff00"
        android:text="Hello, Android" />

</RelativeLayout>
```

实验结果：

- 看到完整的"Hello，Android"，而"Hello，World and Brad"只看到后面部分字符串。
- 没有指定相对位置的组件，默认从左上角开始摆放。
- XML 文件中最先提及的组件在最下面，之后一个一个叠上去。

开始进行相对位置的设定。先将"Hello，World and Brad"放在父容器的正中央，而"Hello，Android"则放在"Hello，World and Brad"上面，并且水平居中。

```xml
<?xml version="1.0" encoding="utf-8"?>
<RelativeLayout xmlns:android="http://schemas.android.com/apk/res/android"
    android:layout_width="match_parent"
    android:layout_height="match_parent" >

    <TextView
        android:id="@+id/tv1"
        android:layout_width="wrap_content"
        android:layout_height="wrap_content"
        android:background="#ffff00"
        android:layout_centerInParent="true"
        android:text="Hello, World and Brad" />

    <TextView
        android:layout_width="wrap_content"
        android:layout_height="wrap_content"
        android:layout_above="@id/tv1"
        android:layout_centerHorizontal="true"
        android:background="#00ff00"
        android:text="Hello, Android" />

</RelativeLayout>
```

呈现以下结果：

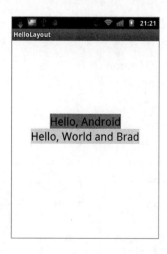

重点说明：

• 虽然呈现结果是"Hello，Android"在上，"Hello，World and Brad"在下，但是在进行版面配置文件时，却是先提及"Hello，World and Brad"，因为其组件的位置条件为在父容器的正中央，只有与父容器有相对位置存在。而"Hello，Android"却必须相依于"Hello，World and Brad"，也就是在其上面，因此必须在之后提及。

• "Hello，World and Brad"的id属性赋予，是因为要给"Hello，Android"组件使用在layout_above属性使用指定。并不一定是在程序开发中会使用，才会赋予id属性。

• 通常在位置中只有相对父容器者会先提及，再来就是相对于已经确定位置之组件的组件位置。

再来一个练习，先看到想要设计的画面。

设计要点：

• 可以分成三大区块，上下中。

• 上列有一个Button在最右边，EditText在最左边，但是EditText也沿着Button的左边，其下方也沿着Button下方。

• 下列中有四个宽度一样的Button。

• 中间是一个TextView，其位置在上列的下面；在下列的上面。

如下实际规划配置：

```
<?xml version="1.0" encoding="utf-8"?>
<RelativeLayout xmlns:android="http://schemas.android.com/apk/res/android"
    android:layout_width="match_parent"
    android:layout_height="match_parent" >

    <RelativeLayout
        android:id="@+id/top"
        android:layout_width="match_parent"
        android:layout_height="wrap_content" >
```

```xml
    <Button
        android:id="@+id/search"
        android:layout_width="wrap_content"
        android:layout_height="wrap_content"
        android:layout_alignParentRight="true"
        android:layout_alignParentTop="true"
        android:text="Search" />

    <EditText
        android:layout_width="match_parent"
        android:layout_height="wrap_content"
        android:layout_alignParentLeft="true"
        android:layout_alignParentTop="true"
        android:layout_toLeftOf="@id/search"
        android:layout_alignBottom="@id/search"
        android:background="#ffff00" />

</RelativeLayout>

<LinearLayout
    android:id="@+id/func"
    android:layout_width="match_parent"
    android:layout_height="wrap_content"
    android:layout_alignParentBottom="true"
    android:orientation="horizontal" >

    <Button
        android:layout_width="match_parent"
        android:layout_height="wrap_content"
        android:layout_weight="1"
        android:text="Add" />

    <Button
        android:layout_width="match_parent"
        android:layout_height="wrap_content"
        android:layout_weight="1"
        android:text="Delete" />

    <Button
        android:layout_width="match_parent"
        android:layout_height="wrap_content"
```

```
            android:layout_weight="1"
            android:text="Config" />
        <Button
            android:layout_width="match_parent"
            android:layout_height="wrap_content"
            android:layout_weight="1"
            android:text="Help" />
    </LinearLayout>
    <TextView
        android:layout_width="match_parent"
        android:layout_height="wrap_content"
        android:layout_above="@id/func"
        android:layout_below="@id/top"
        android:background="#00ff00"
        android:text="Hello, OK"
        />
</RelativeLayout>
```

练习版面配置时，可以为元件加上不同的background属性，用意不是为了好看，而是会清楚该元件所占有的空间有多大。

## 3-4 表格配置TableLayout

顾名思义就是以表格方式呈现内容，常见于呈现特定的表格数据内容，例如个人流水账中的账务内容，或是统计数据、股票信息等。

基本上是以 <TableLayout>…</TableLayout> 为版面配置的方式。

基本架构为<TableLayout> - <TableRow> - <View组件>。如果玩过HTML者，大概都会联想到<table> - <tr> - <td>。

以<TableRow>标示出一列，该列的宽度就是占满<TableLayout>，先来最基本的测试：

```
<?xml version="1.0" encoding="utf-8"?>
<TableLayout xmlns:android="http://schemas.android.com/apk/res/android"
```

```
        android:layout_width="fill_parent"
        android:layout_height="fill_parent" >
    <TableRow>
        <TextView
            android:background="#ffff0000"
            android:text="Data 1" />
        <TextView
            android:background="#ff00ff00"
            android:text="Data 2" />
        <TextView
            android:background="#ff0000ff"
            android:text="Data 3" />
    </TableRow>
</TableLayout>
```

其呈现结果如下图所示。

不是很美观。因为 TextView 预设下是依照内容多寡来决定其占有空间。开始进行修正处理：在 <TableLayout> 中加上属性。

```
android:stretchColumns="0"
```

马上就有不一样的效果呈现，发现是将第 1 个数据域进行扩展，而第 2、3 个数据项，就是依照数据内容决定大小。

再做些修正。

```
android:stretchColumns="0,2"
```

发现是将第 1 个及第 3 个数据域进行平均分配的扩展，而第 2 个数据项，就是依照数据内容决定大小。

通常还可以在<TableRow>之间加上一个<View>类似做出分隔线的效果。

```
<View
    android:layout_width="fill_parent"
    android:background="#ffffff00"
    android:layout_height="2dp"
    />
```

如果特定数据项的位置要留白，则其他数据项最好指定其 layout_column 属性，例如：

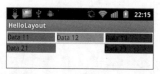

```
<?xml version="1.0" encoding="utf-8"?>
<TableLayout xmlns:android="http://schemas.android.com/apk/res/android"
    android:layout_width="fill_parent"
    android:layout_height="fill_parent"
    android:stretchColumns="0,1,2" >

    <TableRow>

        <TextView
            android:layout_margin="2dp"
            android:background="#ffff0000"
            android:text="Data 11" />

        <TextView
            android:layout_margin="2dp"
            android:background="#ff00ff00"
            android:text="Data 12" />

        <TextView
            android:layout_margin="2dp"
            android:background="#ff0000ff"
            android:text="Data 13" />

    </TableRow>

    <View
        android:layout_width="fill_parent"
```

```
        android:layout_height="2dip"
        android:background="#ffffff00" />
<TableRow>
    <TextView
        android:layout_column="0"
        android:background="#ffff0000"
        android:text="Data 21" />
    <TextView
        android:layout_column="2"
        android:background="#ff0000ff"
        android:text="Data 23" />
</TableRow>
</TableLayout>
```

## 3-5 网格显示GridView

GridView 用来呈现出格子式的配置，最常见的就是相片的陈列。先以早期官方网站的方式来解说，再以不同的变化来更弹性地处理说明。

建立以 GridView 为版面配置的 xml：

```
<?xml version="1.0" encoding="utf-8"?>
<GridView xmlns:android="http://schemas.android.com/apk/res/android"
    android:id="@+id/gv"
    android:layout_width="fill_parent"
    android:layout_height="fill_parent"
    android:columnWidth="90dp"
    android:numColumns="auto_fit"
    android:verticalSpacing="10dp"
    android:horizontalSpacing="10dp"
    android:stretchMode="columnWidth"
    android:gravity="center"
/>
```

开发一个自定义类别继承 BaseAdapter：

```java
private class ImageAdapter extends BaseAdapter {
    private Context mContext;
    public ImageAdapter(Context c) {
        mContext = c;
    }
    public int getCount( ) {
        return mThumbIds.length;
    }
    public Object getItem(int position) {
        return null;
    }
    public long getItemId(int position) {
        return 0;
    }
    public View getView(int position, View convertView, ViewGroup parent) {
        ImageView imageView;
        if (convertView == null) {
            imageView = new ImageView(mContext);
            imageView.setLayoutParams(new GridView.LayoutParams(85, 85));
            imageView.setScaleType(ImageView.ScaleType.CENTER_CROP);
            //imageView.setPadding(8, 8, 8, 8);
        } else {
            imageView = (ImageView) convertView;
        }
        imageView.setImageResource(mThumbIds[position]);
        return imageView;
    }
    private Integer[] mThumbIds = {
            R.drawable.android, R.drawable.apple,
            R.drawable.bells, R.drawable.brick_dig,
            R.drawable.bricks, R.drawable.cup,
            R.drawable.delete, R.drawable.edit,
            R.drawable.enemy1, R.drawable.enemy2,
            R.drawable.eraser, R.drawable.gift,
```

```
            R.drawable.help, R.drawable.mango,
            R.drawable.next, R.drawable.open,
            R.drawable.save, R.drawable.share,
            R.drawable.speed
    };
}
```

回到onCreate( )方法中：

```
private GridView gv;
@Override
protected void onCreate(Bundle savedInstanceState) {
    super.onCreate(savedInstanceState);
    setContentView(R.layout.gridview1);
    gv = (GridView)findViewById(R.id.gv);
    gv.setAdapter(new ImageAdapter(this));
}
```

将会呈现以下效果（前提是也将相对的影像图文件放进res/drawable/子目录下）。

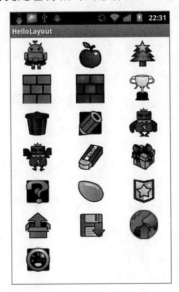

其实GridView的原理观念与ListView类似，因此以下利用在3-1小节的观念原理实作。

先从单一项目的版面配置处理：

```xml
<?xml version="1.0" encoding="utf-8"?>
<LinearLayout xmlns:android="http://schemas.android.com/apk/res/android"
    android:layout_width="wrap_content"
    android:layout_height="wrap_content"
    android:orientation="vertical" >

    <ImageView
        android:id="@+id/gvimg"
         android:layout_width="wrap_content"
         android:layout_height="wrap_content"
         android:src="@drawable/android"/>
    <TextView
        android:id="@+id/gvtv"
        android:layout_width="wrap_content"
        android:layout_height="wrap_content"
        android:layout_gravity="center_horizontal"
        android:text="Android"
        />

</LinearLayout>
```

再回到程序中处理，一样利用SimpleAdapter来进行转换：

```
@Override
protected void onCreate(Bundle savedInstanceState) {
    super.onCreate(savedInstanceState);
    setContentView(R.layout.gridview1);
//    gv = (GridView) findViewById(R.id.gv);
//    gv.setAdapter(new ImageAdapter(this));
    gv = (GridView)findViewById(R.id.gv);

    String[] from = {"gvimg", "gvtv"};
    int[] to = {R.id.gvimg, R.id.gvtv};
    ArrayList<HashMap<String,Object>> data = new ArrayList<HashMap<String,Object>>( );
    HashMap<String,Object> data1 = new HashMap<String,Object>( );
    data1.put("gvimg", R.drawable.android); data1.put("gvtv", "data1");
```

```
data.add(data1);
    HashMap<String,Object> data2 = new HashMap<String,Object>( );
    data2.put("gvimg", R.drawable.apple); data2.put("gvtv", "data2");
data.add(data2);
    HashMap<String,Object> data3 = new HashMap<String,Object>( );
    data3.put("gvimg", R.drawable.mango); data3.put("gvtv", "data3");
data.add(data3);
    HashMap<String,Object> data4 = new HashMap<String,Object>( );
    data4.put("gvimg", R.drawable.bells); data4.put("gvtv", "data4");
data.add(data4);
    HashMap<String,Object> data5 = new HashMap<String,Object>( );
    data5.put("gvimg", R.drawable.cup); data5.put("gvtv", "data5");
data.add(data5);
    gv.setAdapter(new SimpleAdapter(this, data, R.layout.gvtest, from,
to));

}
```

就可以呈现出图文并茂的网格项目：

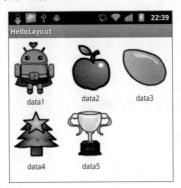

## 3-6 滑动显示ViewFlipper

ViewFlipper通常用来处理在一个Activity中，会有两个以上的ViewGroup，而切换方式可能会搭配手势左右滑动。这样的处理模式意味着一个Activity想要显示画面在移动装置，硬塞下去不好看，所以分成两个以上的画面呈现。

以下实例说明处理方式，假设有三个呈现画面。

（1）第一个画面

```xml
<LinearLayout
    android:layout_width="match_parent"
    android:layout_height="match_parent"
    android:background="#ffffff"
    >
    <TextView
        android:layout_width="match_parent"
        android:layout_height="match_parent"
        android:text="第一个画面" >
    </TextView>
</LinearLayout>
```

（2）第二个画面

```xml
<LinearLayout
    android:layout_width="match_parent"
    android:layout_height="match_parent"
    android:background="#ffff00"
    >
    <TextView
        android:layout_width="match_parent"
        android:layout_height="match_parent"
        android:text="第二个画面" >
    </TextView>
</LinearLayout>
```

（3）第三个画面

```xml
<LinearLayout
    android:layout_width="match_parent"
    android:layout_height="match_parent"
    android:background="#00ff00"
    >
    <TextView
        android:layout_width="match_parent"
```

```
            android:layout_height="match_parent"
            android:text="第三个画面" >
        </TextView>
</LinearLayout>
```

三个画面的背景颜色也是让练习时了解切换的实际状况。

将三个画面直接包进 ViewFlipper 中间。

res/vflipper1.xml

```
<?xml version="1.0" encoding="utf-8"?>
<ViewFlipper xmlns:android="http://schemas.android.com/apk/res/android"
    android:id="@+id/vf"
    android:layout_width="match_parent"
    android:layout_height="match_parent"
    android:background="#000000" >

</ViewFlipper>
```

这样就处理好版面配置的程序了，接下来处理 Activity 的控制。

声明变量：

```
private ViewFlipper vf;
```

取回使用并设定触摸事件：

```
vf = (ViewFlipper)findViewById(R.id.vf);

vf.setOnTouchListener(new OnTouchListener( ) {
    @Override
    public boolean onTouch(View v, MotionEvent event) {
vf.showNext( );
        return super.onTouchEvent(event);
    }
});
```

ViewFlipper 对象实体的 showNext( ) 会直接跳到其下一个子 View 呈现；showPrevios( ) 则刚好相反。所以上面的处理下，用户摸一下就换页，无效果可言。为了达到手势侦测，先建立一个 GestureDetector 对象实体，并处理其事件程序。

声明手势侦测对象变量：

```
private GestureDetector gd;
```

先开发手势侦测事件监听器类别：

```
private class MyGDListener extends GestureDetector.
SimpleOnGestureListener {
    @Override
    public boolean onDown(MotionEvent e) {
          return super.onDown(e);
    }
    @Override
    public boolean onFling(MotionEvent e1, MotionEvent e2, float velocityX,
          float velocityY) {
          return super.onFling(e1, e2, velocityX, velocityY);
    }
}
```

onDown( )方法要Override，并将其传回值设定为true，才会往下onFling( )方法进行侦测手势划过屏幕事件。

- 第三个参数velocityX，表示在 X 轴的划过速度（每秒像素）。
- 往右为正值velocityX。
- 往左为负值velocityX。

所以改写如下：

```
private class MyGDListener extends GestureDetector.
SimpleOnGestureListener {
    @Override
    public boolean onDown(MotionEvent e) {
          return true;
    }
    @Override
    public boolean onFling(MotionEvent e1, MotionEvent e2, float velocityX,
          float velocityY) {
          if (velocityX < -10){
             vf.showNext( );
          }else if(velocityX > 10){
             vf.showPrevious( );
```

```
        }
                return super.onFling(e1, e2, velocityX, velocityY);
        }
}
```

建构出 GestureDetector 对象实体:

```
gd = new GestureDetector(this, new MyGDListener( ));
```

从 ViewFlipper 对象实体的 onTouchEvent( ) 设定传回值,交给 gd.onTouch Event( ):

```
vf.setOnTouchListener(new OnTouchListener( ) {
    @Override
    public boolean onTouch(View v, MotionEvent event) {
            return gd.onTouchEvent(event);
    }
});
```

再来处理动画效果,在 res/anim/ 中建立以下四种动画:
用在 ViewFlipper 的 showNext( ),也就是翻下一页。
从左到右进入:in_leftright.xml。

```xml
<?xml version="1.0" encoding="utf-8"?>
<set xmlns:android="http://schemas.android.com/apk/res/android" >
    <translate
        android:duration="1000"
        android:fromXDelta="-100%"
        android:toXDelta="0" />
</set>
```

- 针对 X 轴的动态效果。
- 进入者之前的的左边为 −100%。
- 进入后为 0 定位显示。

从左到右离开:out_leftright.xml。

```xml
<?xml version="1.0" encoding="utf-8"?>
<set xmlns:android="http://schemas.android.com/apk/res/android" >
    <translate
```

```
        android:duration="1000"
        android:fromXDelta="0"
        android:toXDelta="100%" />
</set>
```

- 出去者之前的左边是 0。
- 出去后为 100%。

用在 ViewFlipper 的 showPrevios( )，也就是回上一页。

从右到左进入：in_rightleft.xml。

```
<?xml version="1.0" encoding="utf-8"?>
<set xmlns:android="http://schemas.android.com/apk/res/android" >
    <translate
        android:duration="1000"
        android:fromXDelta="100%"
        android:toXDelta="0" />
</set>
```

从右到左离开：out_rightleft.xml。

```
<?xml version="1.0" encoding="utf-8"?>
<set xmlns:android="http://schemas.android.com/apk/res/android" >
    <translate
        android:duration="1000"
        android:fromXDelta="0"
        android:toXDelta="-100%" />
</set>
```

再度回到 Activity 中。

声明动画处理对象变量：

```
private Animation in_l2r, out_l2r, in_r2l, out_r2l;
```

加载动画效果：

```
in_l2r = AnimationUtils.loadAnimation(this, R.anim.in_leftright);
out_l2r = AnimationUtils.loadAnimation(this, R.anim.out_leftright);
```

```
in_r2l = AnimationUtils.loadAnimation(this, R.anim.in_rightleft);
out_r2l = AnimationUtils.loadAnimation(this, R.anim.out_rightleft);
```

再度改写onFling( ):

```
public boolean onFling(MotionEvent e1, MotionEvent e2, float velocityX,
    float velocityY) {
    if (velocityX < -10) {
       vf.setInAnimation(in_r2l);
       vf.setOutAnimation(out_r2l);
       vf.showNext( );
    } else if (velocityX > 10) {
          vf.setInAnimation(in_l2r);
          vf.setOutAnimation(out_l2r);
          vf.showPrevious( );
    }
    return super.onFling(e1, e2, velocityX, velocityY);
}
```

OK，相当简单。

# MEMO

# 第4课 对话框与通知事件处理

4-1 AlertDialog 对话框的使用

4-2 自定义对话框（Dialog）与日期时间对话框

4-3 Toast 及自定义 Toast

4-4 进度显示对话框

4-5 通知列处理模式

## 4-1 AlertDialog对话框的使用

对话框对于移动装置而言，可以说是使用频率相当高的项目之一。在用户界面中，通过对话框与用户互动，进而决定程序运行的下一步，或是确认动作的执行等。在Android API中提供了AlertDialog的类别用以快速地处理对话框的开发，android.app.AlertDialog是继承自android.app.Dialog，所以其所建构的对象实体 is-a Dialog，只需要针对开发者自定义部分进行处理即可。

### ■ 4-1-1 建立AlertDialog对象

最常见的方式就是通过AlertDialog的内部类别Builder进行相关的设定，设定之后再调用create( )方法而传回一个AlertDialog对象实体，之后就可以调用AlertDialog对象实体的show( )方法将对话框呈现出来。

- AlertDialog.Builder builder = new AlertDialog.Builder(this);
- builder.setXxx( );
- AlertDialog alert = builder.create( );
- alert.show( )

### ■ 4-1-2 消息对话框

只有提供消息给用户对话框，用户看完之后按下移动装置的返回键即可离开。只需要通过AlertDialog.Builder对象进行设定两项内容：

- setTitle( ): 设定对话框标题。
- setMessage( ): 设定消息内容。
- setCancelable( ): 设定是否允许用户自行关闭（按下返回键）对话框（预设为true）。
- setIcon( ): 设定显示的图标。

因此处理方式非常简单：

```
private void showAlertDialog1( ){
    AlertDialog alert = null;
    AlertDialog.Builder builder = new AlertDialog.Builder(this);
    builder.setTitle("确认对话框");
    builder.setMessage("欢迎使用超级强大功能的App");
    alert = builder.create( );
    alert.show( );
}
```

如果觉得要用户自行按下返回键的动作不恰当，也可以考虑搭配线程的处理，使该消息在指定的时间内呈现，之后自行消失。通常这种模式是强迫用户阅读对话框，一般会搭配周期时间来使对话框自动消失，而用户无法按下返回键使其提前消失。

以下范例的处理将会使对话框出现4s后自动消失，用户无法提前按下返回键使其消失。

```
package tw.brad.android.book.MyAlertDialog;
import java.util.Timer;
import java.util.TimerTask;

import android.app.Activity;
import android.app.AlertDialog;
import android.os.Bundle;
import android.os.Handler;
import android.os.Message;
import android.view.View;
import android.view.View.OnClickListener;
import android.widget.Button;
import android.widget.TextView;

public class MainActivity extends Activity {
    private Button dialog1, dialog2;
    private AlertDialog alert;
    private Timer timer;
    private MyHandler handler;

    @Override
    protected void onCreate(Bundle savedInstanceState) {
```

```java
        super.onCreate(savedInstanceState);
        setContentView(R.layout.activity_main);

        timer = new Timer( );
        handler = new MyHandler( );

        dialog1 = (Button)findViewById(R.id.dialog1);
        dialog1.setOnClickListener(new OnClickListener( ) {
            @Override
            public void onClick(View v) {
                showAlertDialog1( );
            }
        });
    }
    private void showAlertDialog1( ){
        AlertDialog.Builder builder = new AlertDialog.Builder(this);

        builder.setTitle("确认对话框");
        builder.setMessage("欢迎使用超级强大功能的App");
        builder.setCancelable(false);

        alert = builder.create( );
        alert.show( );
        timer.schedule(new CloseDialogTask( ), 4000);
    }
    private class CloseDialogTask extends TimerTask {
        @Override
        public void run( ) {
            handler.sendEmptyMessage(0);
        }
    }

    private class MyHandler extends Handler {
        @Override
        public void handleMessage(Message msg) {
            alert.dismiss( );
        }
    }
}
```

## 4-1-3 确认对话框

与上个对话框处理模式相同，但是用户可以按下确认键或是再多一个取消按键作出确认或是决定，甚至于同时有三种状况的按钮。处理方式就是加上三种按钮的设定，任何一个按钮都会使其对话框自动消失。

Builder 为 AlertDialog.Builder 的对象实体，以下来处理 2 个按钮：（不一定同时三个按钮都实作，视需要而定）。

- setPositiveButton( ) => 右边
- setNegativeButton( ) => 左边
- setNeutralButton( ) => 中间

```java
builder.setPositiveButton("确定(右)", new DialogInterface.OnClickListener( ) {
    @Override
    public void onClick(DialogInterface dialog, int which) {
        // 按下确定后的执行程序
    }
});

builder.setNegativeButton("取消(左)", new DialogInterface.OnClickListener( ) {
    @Override
    public void onClick(DialogInterface dialog, int which) {
        // 按下取消后的执行程序
    }
});

builder.setNeutralButton("再说(中)", new DialogInterface.OnClickListener( ) {
    @Override
    public void onClick(DialogInterface dialog, int which) {
        info.setText("刚刚按下再说");
    }
});
```

当然，也可以只任选一个按钮事件来处理。但是要切记的一点是，代入的第二个参数为实作 DialogInterface.OnClickListener 接口的对象实体，所以是 new DialogInterface.OnClickListener( )，而不是 new View.OnClickListener( )。

使用一个按钮的状况，通常是让用户阅读完毕后自行按下按钮确认，例如显示一份同意书，条文规定等，或是告知目前提供信息；使用两个按钮的状况，通常是请用户当下做出决定，例如是/否，确定/取消，或是登入/重置等。

或是改成询问确认对话框:

```java
private void showAlertDialog2( ){
    AlertDialog.Builder builder = new AlertDialog.Builder(this);

    builder.setTitle("确认对话框");
    builder.setMessage("请输入文件名称");
    builder.setIcon(R.drawable.question);

    filename = new EditText(MainActivity.this);
    builder.setView(filename);

    builder.setPositiveButton("确定", new DialogInterface.OnClickListener( ) {
        @Override
        public void onClick(DialogInterface dialog, int which) {
            info.setText("刚刚按下确定:" + filename.getText( ).toString( ));
        }
    });
    builder.setNeutralButton("再说"new DialogInterface.OnClickListener( ) {
        @Override
        public void onClick(DialogInterface dialog, int which) {
```

```
            info.setText("刚刚按下再说");
        }
    });
    builder.setNegativeButton("取消", new DialogInterface.OnClickListener( ) {
        @Override
        public void onClick(DialogInterface dialog, int which) {
            info.setText("刚刚按下取消");
        }
    });
    builder.setCancelable(false);
    alert = builder.create( );
    alert.show( );
}
```

## 4-1-4 选择式对话框

通常用于四个以上选项的对话框，要求用户在多个选项中做出一个或多个选择。例如以下状况：图像文件案 brad.jpg 已经修改……
・立即保存。
・另存新文件。
・放弃修改。
・删除文件。
处理要点如下：
・通常也会有 setTitle( )。
・但是没有 setMessage( ) => 重要。
・选择项目是通过 setItems( )，传入第一个参数为一个字符串数组为其选项内容，第二个参数为按下选项后的事件处理。

```
// 单一直接选择式
builder.setItems(opts, new DialogInterface.OnClickListener( ) {
    @Override
    public void onClick(DialogInterface dialog, int which) {
        info.setText("您刚刚选择的是: " + opts[which]);
    }
});
```

用户按下特定项目后就直接视为完成选择。

有人会觉得用户可能因为手指触控误差而改成确认处理模式，就必须使用调用 setSingleChoiceItems( )方法来处理，并加上 setXxxButton( )作确认的动作。

```
// 单一确认选择式
builder.setSingleChoiceItems(opts, 0, new DialogInterface.OnClickListener( ) {
    @Override
    public void onClick(DialogInterface dialog, int which) {
        whichItem = which;
    }
});

builder.setPositiveButton("确定", new DialogInterface.OnClickListener( ) {
    @Override
    public void onClick(DialogInterface dialog, int which) {
        info.setText("刚刚按下确定的项目是:" + opts[whichItem]);
    }
});

builder.setNegativeButton("取消", new DialogInterface.OnClickListener( ) {
```

```
    @Override
    public void onClick(DialogInterface dialog, int which) {
        info.setText("刚刚按下取消");
    }
});
```

也可以处理成多选式，就是以调用setMultiChoiceItems( )来处理，但是就必须处理一个以上的setXxxButton( )方法，负责将最后结果作确认或是取消。

```
// 多重选择式
builder.setMultiChoiceItems(opts, checkedItems, new OnMultiChoiceClickListener( ) {
    @Override
    public void onClick(DialogInterface dialog, int which, boolean isChecked) { }
});
builder.setPositiveButton("确定", new DialogInterface.OnClickListener( ) {
    @Override
    public void onClick(DialogInterface dialog, int which) {
        info.setText("刚刚按下确定的项目是:\n");
        for (int i=0; i<opts.length; i++){
            if (checkedItems[i]){
                info.append(opts[i] + "\n");
            }
        }
    }
});
```

```
builder.setNegativeButton("取消", new DialogInterface.OnClickListener( ) {
    @Override
    public void onClick(DialogInterface dialog, int which) {
        info.setText("刚刚按下取消");
});
```

### ■ 4-1-5 进阶选择式对话框

如果觉得前三者的对话框过于单调，可以考虑多点处理动作，以listAdapter来处理会比较有质感。这种形式的对话框与上述的选择式对话框类似，只是将setItems( )改为setAdapter( )来处理，而listAdapter的处理模式与ListView相同。

① 建立一个选项内容的版面。
② 设定Adapter的对应关系。
③ builder.setAdapter( )。

单一选项的版面处理：

```
res/layout/menuitem.xml    <?xml version="1.0" encoding="utf-8"?>
<LinearLayout xmlns:android="http://schemas.android.com/apk/res/android"
    android:layout_width="match_parent"
    android:layout_height="wrap_content"
    android:orientation="horizontal"
    >
    <ImageView
```

```xml
            android:id="@+id/item_img"
            android:layout_width="wrap_content"
            android:layout_height="wrap_content"
        />
    <LinearLayout
        android:layout_width="match_parent"
        android:layout_height="wrap_content"
        android:orientation="vertical"
        <TextView
            android:id="@+id/item_title"
            android:layout_width="match_parent"
            android:layout_height="wrap_content"
            android:textSize="32sp"
            android:textStyle="bold|italic"
        />
        <TextView
            android:id="@+id/item_content"
            android:layout_width="match_parent"
            android:layout_height="wrap_content"
            android:textSize="16sp"
        />
    </LinearLayout>
</LinearLayout>
```

程序处理：

```java
private void showAlertDialog4( ){
    AlertDialog.Builder builder = new AlertDialog.Builder(this);
    builder.setTitle("您的移动装置操作系统为何?");
    builder.setIcon(R.drawable.opts);
    data = new ArrayList( );
    HashMap<String,Object> item0 = new HashMap( );
    item0.put(from[0], R.drawable.android);
    item0.put(from[1], opts[0]);
```

```java
item0.put(from[2], "就是你现在正在学习的东西呀");
data.add(item0);

HashMap<String,Object> item1 = new HashMap( );
item1.put(from[0], R.drawable.apple);
item1.put(from[1], opts[1]);
item1.put(from[2], "另一个阵营的好东西");
data.add(item1);

HashMap<String,Object> item2 = new HashMap( );
item2.put(from[0], R.drawable.windows);
item2.put(from[1], opts[2]);
item2.put(from[2], "PC市场上无人不知无人不晓");
data.add(item2);

HashMap<String,Object> item3 = new HashMap( );
item3.put(from[0], R.drawable.other);
item3.put(from[1], opts[3]);
item3.put(from[2], "嗯...");
data.add(item3);

SimpleAdapter adapter = new SimpleAdapter(
        this, data,
        R.layout.menuitem,
builder.setAdapter(adapter, new DialogInterface.OnClickListener( ) {
    @Override
    public void onClick(DialogInterface dialog, int which) {
        info.setText("您刚刚选择的是：" + opts[which]);
    }
});

alert = builder.create( );
alert.show( );
}
```

## 4-2 自定义对话框（Dialog）与日期时间对话框

### ■ 4-2-1 自定义对话框

　　自定义对话框通常应用于对话框内容，会提供不同于单纯选项的互动输入状况，或是想要提供美观性更高的用户界面。处理的对象为以 android.app.Dialog 类别对象为主，版面配置通常会以 xml 格式的 layout 文件为呈现内容，再从程序中建构出 Dialog 对象实体，两者进行结合使用。所以连用户界面的相关确认按钮、相关组件及事件处理，都不像上一小节中的 AlertDialog 中只需要 Override 其特定事件倾听事件即可，甚至于连关闭对话框的动作程序都不是自动处理，而优点就是版面配置的弹性比较大。

　　以下的注册对话框就是必须以自定义方式来处理：

　　先设计出希望呈现的版面配置：

```
res/layout/newreg.xml
<?xml version="1.0" encoding="utf-8"?>
<LinearLayout xmlns:android="http://schemas.android.com/apk/res/android"
    android:layout_width="match_parent"
    android:layout_height="match_parent"
    android:orientation="vertical"
    >
    <TextView
        android:layout_width="match_parent"
```

```xml
        android:layout_height="wrap_content"
        android:textSize="18sp"
        android:text="拉拉网账号"
     />
<EditText
       android:id="@+id/reg_username"
        android:layout_width="match_parent"
        android:layout_height="wrap_content"
        android:singleLine="true"
     />
<TextView
        android:layout_width="match_parent"
        android:layout_height="wrap_content"
        android:textSize="18sp"
        android:text="拉拉网昵称"
     />
<EditText
        android:id="@+id/reg_nickname"
        android:layout_width="match_parent"
        android:layout_height="wrap_content"
        android:singleLine="true"
     />
<TextView
        android:layout_width="match_parent"
        android:layout_height="wrap_content"
        android:textSize="18sp"
        android:text="使用的密码"
     />
<EditText
        android:id="@+id/reg_passwd"
        android:layout_width="match_parent"
        android:layout_height="wrap_content"
        android:password="true"
        android:singleLine="true"
     />
<TextView
        android:layout_width="match_parent"
```

```
        android:layout_height="wrap_content"
        android:textSize="18sp"
        android:text="再次输入以确认"
    />
<EditText
    android:id="@+id/reg_passwd2"
     android:layout_width="match_parent"
     android:layout_height="wrap_content"
     android:password="true"
     android:singleLine="true"
    />
<TextView
     android:layout_width="match_parent"
     android:layout_height="wrap_content"
     android:textSize="18sp"
     android:text="手机号码"
    />
<EditText
     android:id="@+id/reg_phonenum"
     android:layout_width="match_parent"
     android:layout_height="wrap_content"
     android:singleLine="true"
     android:numeric="integer"
    />

<LinearLayout
     android:layout_width="match_parent"
     android:layout_height="wrap_content"
     android:orientation="horizontal"
     >
    <Button
         android:id="@+id/reg_ok"
         android:layout_width="match_parent"
         android:layout_height="wrap_content"
         android:layout_weight="1"
         android:text="申请"
     />
```

```xml
        <Button
            android:id="@+id/reg_cancel"
            android:layout_width="match_parent"
            android:layout_height="wrap_content"
            android:layout_weight="1"
            android:text="取消"
            />
    </LinearLayout>
</LinearLayout>
```

在样式上也设定了背景颜色相关:

```
values/styles.xml 中
<style name="dialogStyle" parent="@android:style/Theme.Dialog">
    <item name="android:background">#449900</item>
</style>
```

而在程序中先将 Dialog 对象建构出来:

```
dialog_newreg = new Dialog(this, R.style.dialogStyle);
```

并设定为上述的版面配置:

```
dialog_newreg.setContentView(R.layout.newreg);
```

进行与 AlertDialog 相同的基本设定:

```
dialog_newreg.setTitle("拉拉网 - 新用户注册");
dialog_newreg.setCancelable(false);
```

而在版面中,用户互动输入相关组件的处理,则必须由 Dialog 对象实体来调用 findViewById( ) 方法传回。

```
eUsername = (EditText)dialog_newreg.findViewById(R.id.reg_username);
eNickname = (EditText)dialog_newreg.findViewById(R.id.reg_nickname);
ePasswd = (EditText)dialog_newreg.findViewById(R.id.reg_passwd);
ePasswd2 = (EditText)dialog_newreg.findViewById(R.id.reg_passwd2);
ePhonenum = (EditText)dialog_newreg.findViewById(R.id.reg_phonenum);
```

```
Button reg_ok, reg_cancel;
reg_ok = (Button)dialog_newreg.findViewById(R.id.reg_ok);
reg_cancel = (Button)dialog_newreg.findViewById(R.id.reg_cancel);
```

最后两个Button的处理与AlertDialog中的Button不一样的是一般Button的OnClick事件，而不是使用DialogInterface.OnClick事件。

```
reg_ok.setOnClickListener(new OnClickListener( ) {
    @Override
    public void onClick(View v) {
    }
reg_cancel.setOnClickListener(new OnClickListener( ) {
    @Override
    public void onClick(View v) {
    }
});
```

用户按下Button之后并不会使Dialog对象自动消失，除非调用Dialog对象之dismiss( )方法，才会使其消失。

```
dialog_newreg.dismiss( );
```

当然，最后设定完成后，必须调用Dialog对象之show( )方法才能使其呈现出来。

### ■ 4-2-2　日期选择对话框

android.app.DatePickerDialog类别对象用来产生出选择日期的选取对话框，提供用户互动友善的选取日期。

建构对象方式通常会需要设定初始的年月日数据，可以通过java.util.Calendar类别对象取得目前的年月日数据。

```
int year, month, day;
Calendar cal = Calendar.getInstance( );
year = cal.get(Calendar.YEAR);
month = cal.get(Calendar.MONTH);
day = cal.get(Calendar.DAY_OF_MONTH);
```

接着就可建构出对象实体:

```
dpd = new DatePickerDialog(this, DatePickerDialog.THEME_HOLO_DARK,
    new OnDateSetListener( ) {
        @Override
        public void onDateSet(DatePicker view, int year, int monthOfYear,
                                                int dayOfMonth) {
            info.setText(year + "/" + monthOfYear + "/" + dayOfMonth);
        }
    },
    year, month, day);
```

参数说明:
- 目前的Context。
- 呈现的样式风格。

DatePickerDialog.THEME_TRADITIONAL
DatePickerDialog.THEME_HOLO_DARK
DatePickerDialog.THEME_HOLO_LIGHT
DatePickerDialog.THEME_DEVICE_DEFAULT_DARK
DatePickerDialog.THEME_DEVICE_DEFAULT_LIGHT

- 变更的事件处理器。
- 初始的年份值。
- 初始的月份值。
- 初始的日期值。

再来就是进行一般设定:

```
dpd.setTitle("设定日期");
dpd.setButton(DatePickerDialog.BUTTON_POSITIVE, "确定", new
DialogInterface.OnClickListener( ) {
    @Override
    public void onClick(DialogInterface dialog, int which) {
        dpd.dismiss( );
    }
}
dpd.setButton(DatePickerDialog.BUTTON_NEGATIVE, "取消", new
DialogInterface.OnClickListener( ) {
    @Override
```

```
    public void onClick(DialogInterface dialog, int which) {
        dpd.dismiss( );
    }
});
```

最后记得将其呈现出来。

```
dpd.show( );
```

### ■ 4-2-3　时间选择对话框

　　android.app.TimePickerDialog类别对象用来产生出选择时间的选取对话框，提供用户互动友善的选取时间。

　　建构对象方式通常会需要设定初始的时、分数据，可以通过java.util.Calendar类别对象取得目前的时分数据。

```
int hour, minute;
Calendar cal = Calendar.getInstance( );
hour = cal.get(Calendar.HOUR_OF_DAY);
minute = cal.get(Calendar.MINUTE);
```

　　接着就可建构出对象实体：

```
tpd = new TimePickerDialog(this,TimePickerDialog.THEME_HOLO_DARK,
    new OnTimeSetListener( ) {
        @Override
        public void onTimeSet(TimePicker view, int hourOfDay, int minute) {
```

```
                    info.setText(hourOfDay + ":" + minute);
        }
    },
    hour, minute, true);
```

参数说明：
① 目前的Context。
② 呈现的样式风格。
- TimePickerDialog.THEME_TRADITIONAL。
- TimePickerDialog.THEME_HOLO_DARK。
- TimePickerDialog.THEME_HOLO_LIGHT。
- TimePickerDialog.THEME_DEVICE_DEFAULT_DARK。
- TimePickerDialog.THEME_DEVICE_DEFAULT_LIGHT。
③ 变更的事件处理器。
④ 初始的小时值。
⑤ 初始的分钟值。
⑥ 是否为24小时制（boolean）。

再来就是进行一般设定：

```
tpd.setTitle("设定时间");
tpd.setButton(DatePickerDialog.BUTTON_POSITIVE, "确定", new
DialogInterface.OnClickListener( ) {
    @Override
    public void onClick(DialogInterface dialog, int which) {
        tpd.dismiss( );
    }
});
tpd.setButton(DatePickerDialog.BUTTON_NEGATIVE, "取消", new
DialogInterface.OnClickListener( ) {
    @Override
    public void onClick(DialogInterface dialog, int which) {
        tpd.dismiss( );
    }
});
```

最后记得将其呈现出来。

```
tpd.show( );
```

## 4-3 Toast及自定义Toast

android.widget.Toast 就像是刚从烤面包机弹出来的土司一样,突然从屏幕上弹出一段消息。不需要去特别作任何操作,只是提供消息而已,稍候一下就会自动消失。

### ■ 4-3-1 一般的Toast

最简单的处理方式,就是直接调用Toast的static方法makeText( ),传递三个参数:
• Context。
• 显示字符串。
• 呈现停留时间。
Toast.LENGTH_LONG
Toast.LENGTH_SHORT
然后直接调用show( )方法呈现出来。

```
Toast.makeText(this, "大家好,偶素Brad", Toast.LENGTH_LONG).show( )
```

或是设定其显示的位置,调用其setGravity( )方法,传递三个参数:
• Gravity:(举例)。
Gravity.CENTER: 屏幕正中央。
Gravity.FILL: 填满屏幕。
Gravity.FILL_HORIZONTAL+Gravity.TOP: 屏幕上方水平填满。
• X轴偏移。
• Y轴偏移。

如下处理范例：

```
Toast tst1 = Toast.makeText(this, "大家好,偶素Brad", Toast.LENGTH_LONG);
tst1.setGravity(Gravity.FILL_HORIZONTAL+Gravity.CENTER_VERTICAL, 0, 0);
tst1.show( );
```

### 4-3-2 自定义Toast

而自定义的Toast的弹性就比较大，整个版面内容可以自行规划设计，这样可以使开发项目App的整体风格较为一致，经常看到许多App在版面设计上非常用心，甚至于画面精致的游戏，结果当出现上述一般的Toast来呈现即时消息时，整个感觉就非常差了，不妨考虑来撰写个自定义版面的Toast呈现方法来处理，会比较有整体的质感表现。

先来设计自定义的版面配置：

res/layout/toast_view.xml

```xml
<?xml version="1.0" encoding="utf-8"?>
<LinearLayout xmlns:android="http://schemas.android.com/apk/res/android"
    android:id="@+id/toast_layout"
    android:layout_width="match_parent"
    android:layout_height="wrap_content"
    android:orientation="horizontal"
    android:background="@drawable/gold_banner"
    >
    <ImageView
        android:id="@+id/toast_icon"
```

```xml
        android:layout_width="wrap_content"
        android:layout_height="wrap_content"
        android:src="@drawable/sms"
        android:layout_margin="12dp"
    />
    <LinearLayout
        android:layout_width="match_parent"
        android:layout_height="wrap_content"
        android:orientation="vertical"
        android:layout_margin="12dp"
    >
        <TextView
            android:id="@+id/toast_title"
            android:layout_width="match_parent"
            android:layout_height="wrap_content"
            android:textSize="24sp"
            android:textStyle="bold"
            android:textColor="#00ff00"
        />
        <TextView
            android:id="@+id/toast_content"
            android:layout_width="match_parent"
            android:layout_height="wrap_content"
            android:textSize="18sp"
            android:textStyle="bold|italic"
            android:textColor="#ffffff"
        />
    </LinearLayout>
</LinearLayout>
```

呈现一个横列信息，左边为一个图标，右边分成上下表现出一个标题与内容。

回到程序中来开发一个方法，可以接收两个参数，分别就是标题字符串与内容字符串，当然，读者也可以将图标同时变更设定处理。

处理重点如下：

- 调用 getLayoutInflater( ) 将会传回一个 LayoutInflater 对象实体。
- 执行该 LayoutInflater 对象实体的 inflater( ) 方法，传入两个参数自定义版面的资源位置 (R.layout.toast_view)。

自定义版面中的ViewGroup对象。
传回该版面的View对象实体。
- 再由View对象实体取得标题与内容的TextView对象实体。
- 设定标题与内容的TextView对象实体所呈现的文字内容。
- 建立Toast对象实体。
- 调用setView( )传入之前的View对象实体。
- 其他处理方式与一般Toast一样。

如下：

```
private void showToast2( ){
    showMyToast("重要讯息", "恭喜您中了大大大乐透");
}
private void showMyToast(String title, String content){
    LayoutInflater inflater = getLayoutInflater( );
    View layout = inflater.inflate(R.layout.toast_view,
                    (ViewGroup) findViewById(R.id.toast_layout));
    TextView toast_title=(TextView) layout.findViewById(R.id.toast_title);
    TextView toast_content=(TextView) layout.findViewById(R.id.toast_content);

    toast_title.setText(title);
    toast_content.setText(content);

    Toast toast = new Toast(getApplicationContext( ));
    toast.setGravity(Gravity.CENTER_VERTICAL, 0, 0);
    toast.setDuration(Toast.LENGTH_LONG);
    toast.setView(layout);
    toast.show( );
}
```

## 4-4 进度显示对话框

进度显示对话框通常是用在当App执行中，可能无法立即执行完毕，而必须请用户稍候得以继续，例如正在下载因特网相关的文件时；或是当App正在加载较大的程序时，告知用户目前的现况，以免用户以为App发生非预期状态（死机），例如App在首次加载执行时，可能必须花较多的时间进行相关资源加载程序。

而进度对话框一般而言可以分成两种，一种是可以掌握运行程序的进度，例如一开始就知道要下载的文件大小，而进行文件传输串流过程中也能够取得目前已经下载的大小，此时可以考虑以呈现目前执行进度的对话框较为恰当。

另一种是无法掌握整体的执行进度相关数据，而只能知道执行的结束点，则只好提供动态循环运转的对话框，让用户知道目前仍然在执行中。

### ■ ProgressDialog

ProgressDialog的建构方式是直接调用该类别所提供的建构式，笔者最常使用的如下：

```
new ProgressDialog(this, ProgressDialog.THEME_HOLO_DARK);
```

传递两个参数：
① Context。
② 呈现风格。
- ProgressDialog.THEME_DEVICE_DEFAULT_DARK(API Level 14)
- ProgressDialog.THEME_DEVICE_DEFAULT_LIGHT(API Level 14)
- ProgressDialog.THEME_DEVICE_HOLO_DARK(API Level 11)
- ProgressDialog.THEME_DEVICE_HOLO_LIGHT(API Level 11)
- ProgressDialog.THEME_DEVICE_TRADITIONAL(API Level 11)

决定是否为进度可以掌控的模式，若无法掌握进度，则调用setProgress Style( )方法时传递ProgressDialog.STYLE_SPINNER。

```
.setProgressStyle(ProgressDialog.STYLE_SPINNER);
```

通常会搭配消息的呈现，则调用setMessage( )方法，传递消息字符串参数进去。这样就算是事先准备工作完成。

以下以一个单纯线程为例，当然，通常也会将因特网相关处理程序放在线程中处理，算是相当常见的处理模式。而线程中必须通过android.os.Handler来进行前台组件的改动，因此先来定义一个自定义的Handler类别。

```java
private class MyHandler extends Handler {
    @Override
    public void handleMessage(Message msg) {
        super.handleMessage(msg);
        switch(msg.what){
            case 1:
                pDialog1.show( );
                break;
            case 2:
                if (pDialog1.isShowing( )){
                    pDialog1.dismiss( );
                }
                break;
        }
    }
}
```

线程类别（handler 为上述之 MyHandler 类别对象）：

```java
private class MyTask1 extends Thread {
    @Override
    public void run( ) {
        handler.sendEmptyMessage(1);
        for (int i=0; i<100; i++){
            try {
                Thread.sleep(100);
            } catch (InterruptedException e) {
                e.printStackTrace( );
            }
        }
        handler.sendEmptyMessage(2);
    }
}
```

当线程开始执行的时候，handler 对象会送值为 1 的 which 值，并进行 ProgressDialog 对象的 show( ) 方法开始呈现……

若可以掌握进度,则调用setProgressStyle( )方法时传递ProgressDialog.STYLE_HORIZONTAL。

```
.setProgressStyle(ProgressDialog.STYLE_HORIZONTAL);
```

另外撰写一个执行绪,与上个范例不同之处是在执行中随时将i值包在Message对象中的Bundle,并由handler对象传送处理。

```
private class MyTask2 extends Thread {
    @Override
    public void run( ) {
        handler.sendEmptyMessage(3);
        for (int i = 0; i < 100; i++) {
            try {
                Message msg = new Message( );
                msg.what = 5;
                Bundle data = new Bundle( );
                data.putInt("progress", i);
                msg.setData(data);
                handler.sendMessage(msg);
                Thread.sleep(100);
            } catch (InterruptedException e) {
                e.printStackTrace( );
            }
        }
        handler.sendEmptyMessage(4);
    }
}
```

改写上述的MyHandler类别定义:

```
case 3:
    pDialog2.setMax(100);
```

```
                pDialog2.show( );
                break;
        case 4:
            }if (pDialog2.isShowing( )) {
                break;
        case 5:
                pDialog2.dismiss( );
            break;
            if (pDialog2.isShowing( )) {
pDialog2.setProgress(msg.getData( ).getInt("progress"));
            }
                break;
    }
```

即可呈现如下结果：

## 4-5 通知列处理模式

在移动装置的通知列通常用来提醒用户发生了特定的事件通知，在发出通知的当时会有特殊的字符串显示，并可以带有其他提示音、振动或是闪光强化通知的效果，达到提醒重要通知的目的。大部分发出通知的状况通常是在App的后台中处理，当用户拉下通知列后，触摸该通知事项时，可以将用户带回App中特定的前台页面。

### ■ 4-5-1 版本差异

通知基本的处理程序大致上有几个不同API Level的处理模式。
- API Level 11之前。
- API Level 11之后。
- API Level 16+。

先来了解共同的处理观念。

① Notification 的处理方式必须通过 NotificationManager 对象负责管理。
② 通过调用 getSystemService（NOTIFICATION_SERVICE）取得。
③ 建构 Notification 对象实体。
④ 建构通知启动的 Intent 对象实体。
⑤ 以 Intent 对象实体建构出 PendingIntent 对象实体。
⑥ 由 NotificationManager 调用 notify( ) 方法，将 Notification 对象实体发出通知。

## ■ 4-5-2　API Level 11 之前

```
private void createNotification( ){
    // 取得 NotificationManager 对象
    NotificationManager mNotificationManager =
            (NotificationManager)getSystemService(NOTIFICATION_SERVICE);
    // 设定通知的图标，通知提示，通知显示的时间
    Notification notification = new Notification(R.drawable.tab10, "重要通知",
                        System.currentTimeMillis( ));
    // 通知启动的 Intent 对象
    Intent notificationIntent = new Intent(this, Page2.class);
    notificationIntent.putExtra("which", 4);   // 传递数据过去
    // 以 Intent 来取得 PendingIntent 对象
    PendingIntent contentIntent = PendingIntent.getActivity(this, 0,
        notificationIntent, 0);
    notification.setLatestEventInfo(this, "期末考快被当""期末考如未达 90
        分，总成绩将会被当", contentIntent);
    // 发出通知
    mNotificationManager.notify(1, notification);
}
```

## ■ 4-5-3　API Level 11+

本书重点放在 API Level 11+ 及 API Level 16+，假设发出如下图的通知：

先开发设计一个 Activity，用来当产生通知事件之后，用户点按后所出现的画面。画面如下：

activity_my_notice_page.xml

```xml
<LinearLayout xmlns:android="http://schemas.android.com/apk/res/android"
    android:layout_width="match_parent"
    android:layout_height="match_parent"
    android:orientation="vertical"
    >

    <TextView
        android:layout_width="wrap_content"
        android:layout_height="wrap_content"
        android:text="Notice Page" />

</LinearLayout>
```

以及对应的 Activity：

MyNoticePage.java

```java
package tw.brad.android.book.MyNotification;

import android.os.Bundle;
import android.app.Activity;
import android.view.Menu;

public class MyNoticePage extends Activity {
    @Override
    protected void onCreate(Bundle savedInstanceState) {
        super.onCreate(savedInstanceState);
        setContentView(R.layout.activity_my_notice_page);
    }
}
```

而回到主程序中，产生一个 Intent，用来指定跳到该 Activity。

```java
Intent intent = new Intent(this, MyNoticePage.class);
```

通过 TaskStackBuilder 来产生一个 PendingIntent 对象实体。

```java
TaskStackBuilder stackBuilder = TaskStackBuilder.create(this);
stackBuilder.addParentStack(MyNoticePage.class);
stackBuilder.addNextIntent(intent);
```

```
PendingIntent pending = stackBuilder.getPendingIntent(0,PendingIntent.
FLAG_UPDATE_CURRENT);
```

参数说明:
① 本 Activity 对象实体。
② RequestCode,用来辨别调用 Intent。
③ 前述的 Intent 对象实体。
④ Flag。
- PendingIntetn.FLAG_ONE_SHOT。
- PendingIntetn.FLAG_NO_CREATE。
- PendingIntetn.FLAG_CANCEL_CURRENT。
- PendingIntetn.FLAG_UPDATE_CURRENT。

建构 Notification.Builder 对象实体

```
Notification.Builder builder = new Notification.Builder(this);
```

开始进行设定:

```
builder.setTicker("重要的通知");
builder.setAutoCancel(true);
builder.setSmallIcon(R.drawable.ic_launcher);
builder.setContentInfo("Info");
builder.setContentText("text");
builder.setContentTitle("title");
builder.setContentIntent(pending);
builder.setWhen(System.currentTimeMillis( ) + 10000); // 延迟10s
builder.setContentIntent(pending);
```

进行建立 Notification 对象实体:
API Level 11+(也即 Android 3.0 以上)。

```
Notification notification = builder.getNotification( );
```

若是 API Level 16+(也即 Android 4.1.2 以上)。

```
Notification notification = builder.build( );
```

最后,以 getSystemService( ) 调用出 NotificationManager 对象实体,以发出通知。

```
NotificationManager mgr = (NotificationManager)
getSystemService(NOTIFICATION_SERVICE);
mgr.notify(0, notification);
```

### ■ 4-5-4　应用场合

　　以上是基本处理方式，但是许多状况都会与Service或是后台线程搭配使用。当App已经从前台执行状态离开，而后台的Service正在执行周期任务，例如每5min联机网站服务器，检查是否有更新数据。如果有更新变动数据内容，马上传递到移动装置，此时也发出通知，引导用户点击通知内容，直接带到指定Activity来呈现更新数据。

# 05
Chapter

## 第5课　进阶程序运行原理与应用

5-1　多重线程 Thread

5-2　定时及周期任务（Timer & TimerTask）

5-3　同步任务 AsyncTask

5-4　倒数定时器

## 5-1 多重线程Thread

在一个App的执行期间，当有多样任务工作可能会同时进行的时候，就会使用到多重线程的处理方式。而多重线程的原理是将要进行的程序片段，放在执行后台中处理，而前台为主程序，也就是主线程。前台只能有一个，而后台线程可以有多个同时进行。

譬如在一个游戏App中，主角是炸弹超人的线程，而在关卡中还有三个坏蛋的线程，分别在后台进行位移运算，并不会因为炸弹超人行进中，而所有坏蛋就停止其行进的动作，应该是各自有不同的线程对象分别处理各自的不同情境状态。或是在打砖块游戏中，玩家拿到宝物后开始发射子弹，而每个子弹也都是个别独立的线程对象来处理，也不会影响目前正在碰撞中的球体行进。

### ■ 5-1-1 开发重点观念

很常见的处理方式就是以内部类别来开发，这样会比较容易存取外部类别对象的成员。

① 自定义类别继承java.lang.Thread。
② Override其run( )。
③ 将该线程生命周期的程序定义撰写在run( )方法中。

```
private class MyThread1 extends Thread {
    @Override
    public void run( ) {
        for (int i=0; i<10; i++){
            Log.i("brad", "i = " + i);
            try {
                Thread.sleep(200);
            } catch (InterruptedException e) {
                e.printStackTrace( );
            }
        }
    }
}
```

上述程序片段为一个内部类别的线程类别，生命周期执行一段 for 循环，每一循环执行 Log 显示变量 i 值，执行后使其调用 Thread 的 static 方法 sleep( )，传入参数为 1/1000s 为单位的值，让该线程暂停至少 0.2s(值为 200)，再继续下一个循环。

先将该物件建构：

```
MyThread1 t1 = new MyThread1( );
```

如果只是以一般对象方法调用 run( )，则并不具有线程的生命周期特征。而必须调用该对象的 start( ) 才会表现出生命周期的特征，以执行 run( ) 方法中所定义的程序区块。

不具生命周期的调用：

```
t1.run( );
```

具有生命周期的调用：

```
t1.start( );
```

虽然具有线程生命周期特征，但是只能调用一次，当 run( ) 方法中的程序区块执行完毕后，就结束其线程的生命周期，如果重复调用 start( )，则将会在第二次调用执行后抛出 RuntimeException。

执行之后会在 LogCat 中看到状况如下：

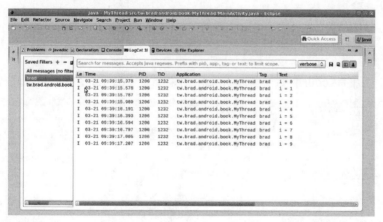

如果同时启动两个相同类别的线程对象来做实验，并改写线程对象的建构式，传递 1 个字符串属性。

```
private class MyThread1 extends Thread {
    String name;
    MyThread1(String n){
```

```
            name = n;
    }
    @Override
    public void run( ) {
        for (int i = 0; i < 10; i++) {
            Log.i("brad", name + ": i = " + i);
            try {
                Thread.sleep(200);
            } catch (InterruptedException e) {
                e.printStackTrace( );
            }
        }
    }
}
```

接着建构出两个对象实体，并分别调用start( )方法：

```
private void doThread1( ) {
    MyThread1 t1 = new MyThread1("A");
    MyThread1 t2 = new MyThread1("B");
    t1.start( );
    t2.start( );
}
```

执行之后的状况可能如下（读者实作结果不一定会是完全相同的顺序性）：

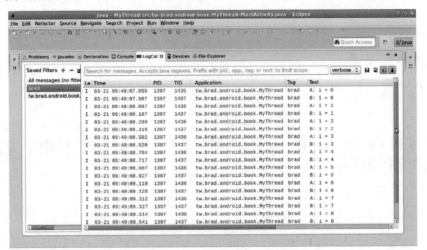

## 5-1-2 存取View组件

如果在run( )方法中存取View组件,例如调用TextView对象实体的setText( )方法。

```
private class MyThread1 extends Thread {
    String name;
    MyThread1(String n) {
        name = n;
    }
    @Override
    public void run( ) {
        for (int i = 0; i < 10; i++) {
            Log.i("brad", name + ": i = " + i);
            msg.setText(name + ": i = " + i);
            try {
                Thread.sleep(200);
            } catch (InterruptedException e) {
                e.printStackTrace( );
            }
        }
    }
}
```

上例中的msg为一个TextView对象。

这样的状况下,并不会发生编译失败,但是会在执行时期出现RuntimeException。

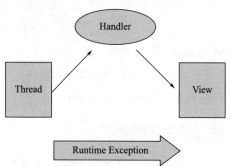

多重线程在Java Appliaction中已经是非常常见的运用,在Android中的使用方式是完全一样的,唯一要特别注意的是,在线程中无法直接对View组件作任何操作,虽然在编译阶段没有任何语法错误,却会在执行时期抛出Running Exception。

因此可以通过继承Android.os.Handler类别的子类别对象来处理。

```java
private class MyHandler extends Handler {
    @Override
    public void handleMessage(Message message) {
            msg.setText(t1.name + ": i = " + message.what);
    }
}
```

- t1 为上例中 MyThread1 线程类别的对象实体。
- msg 为前台画面中的 TextView 类别的对象实体。
- message 为传递进来的 Message 对象实体。

而在上例的线程中修改如下：

```java
@Override
public void run( ) {
    for (int i = 0; i < 10; i++) {
        Log.i("brad", name + ": i = " + i);
        //msg.setText(name + ": i = " + i);
        handler.sendEmptyMessage(i);
        try {
            Thread.sleep(200);
        } catch (InterruptedException e) {
            e.printStackTrace( );
        }
    }
}
```

- handler 为上文中 MyHandler 类别对象。
- 调用 MyHandler 类别对象方法 sendEmptyMessage( )，传递参数 i 值。
- 则将会触发 handler 对象中的 handleMessage( ) 方法。

### ■ 5-1-3 提早结束线程的生命周期

当线程生命周期尚未执行完毕时，可能会因其他因素如用户取消任务执行或是特定事件触发，而必须提早结束该线程生命周期，则可以使用以下模式进行开发。

- 在 run( ) 方法中以 try...catch 捕捉 InterruptedException。
- 在 catch 程序区块中提早结束，例如循环结构的 break 叙述句或是直接 return。

- 而触发InterruptedException方式就是调用该对象实体的interrupt( )将上述范例MyThread1类别中的run( )修改如下：

```
for (int i = 0; i < 100; i++) {
    Log.i("brad", name + ": i = " + i);
    //msg.setText(name + ": i = " + i);
    handler.sendEmptyMessage(i);
    try {
            Thread.sleep(200);
    } catch (InterruptedException e) {
            break;
    }
}
```

当线程对象t1调用了interrupt( )，就会提早脱离循环结构，而结束执行run( )，也就是提早结束其生命周期的表现了。

```
private void stopThread1( ){
    if (t1 != null && t1.isAlive( )){
            t1.interrupt( );
    }
}
```

### ■ 5-1-4 另外一种开发方式

除了上述的自订类别继承自java.lang.Thread外，也常被讨论的方式就是实作（implements）java.lang.Runnable界面的自定义类别。这样处理线程的优点在于该自定义类别尚可继承其他父类别，而相较于上述的继承java.lang.Thread的方式，就无法继承其他父类别。

定义类别方式：

```
private class MyRunnable1 implements Runnable {
    String name;

    MyRunnable1(String n) {
            name = n;
    }

    @Override
```

```
    public void run( ) {
        for (int i = 0; i < 100; i++) {
            Log.i("brad", name + ": i = " + i);
            //msg.setText(name + ": i = " + i);
            handler.sendEmptyMessage(i);
            try {
                Thread.sleep(200);
            } catch (InterruptedException e) {
                break;
            }
        }
    }
}
```

而启动执行生命周期如下：

```
private void doRunnable1( ){
    MyRunnable1 mr = new MyRunnable1("C");
    Thread t3 = new Thread(mr);
    t3.start( );
}
```

重要观念如下：

- 一个Runnable对象实体并非线程对象。
- 线程对象为Thread，也就是上例中的t3。
- t3线程对象是传入一个Runnable对象实体来建构的。
- 所以是由t3调用start( )方法来启动线程的生命周期。
- 执行过程则是表现定义在Runnable对象的run( )方法中。

> **Note** Thread.sleep( )方法的时间值，只是表示该线程的暂停休眠时间，并不代表恢复休眠时间后就马上执行，也就是说时间的精准度不够，通常是大于指定的暂停休眠时间，因此通常用于确保任务执行的完成度，例如从网络提取文件或是资料的程序。如果要求较高的时间精准度，则建议使用下一小节的Timer来处理较佳。

## 5-2 定时及周期任务（Timer & TimerTask）

在App中处理时间周期相关的部分，笔者通常会使用java.util.Timer类别对象来运用。 凡是与定时或是特定周期循环的任务，都可以以一个Timer对象来总管整个所有不同的时间相关任务。而可以被交给Timer对象管理的任务，则必须是java.util.TimerTask类别的对象，但是TimerTask为抽象类别，因此会以自定义类别继承TimerTask类别，并override其run( )来实作定义任务的程序区块。

通常处理模式：
① 建构Timer对象实体。
② 自定义类别继承TimerTask。
③ Override其public void run( )方法。
④ 开发定义任务程序区块。
⑤ 建构出TimerTask对象实体。
⑥ Timer对象调用schedule( )方法，传入TimerTask对象实体，及时间周期相关参数。

与Thread处理模式不同之处在于时间的管控是由Timer对象负责，而且只负责时间任务控制而已；而任务内容就完全不与时间相关，单纯地开发撰写工作任务的内容即可。

建构Timer对象实体：

```
Timer timer = new Timer( );
```

假设开发一个一开始就会有一个TextView呈现移动装置上目前时间的分秒，所以也会需要有一个Handler对象实体来处理前台组件的存取动作。

```
private class MyHandler extends Handler {
    @Override
    public void handleMessage(Message msg) {
        if (msg.what == 0){
            Calendar now = Calendar.getInstance( );
            now.setTime(new Date(System.currentTimeMillis( )));
            msg2.setText(now.get(Calendar.MINUTE) + ":" + now.
                get(Calendar.SECOND));
        }else {
        }
    }
}
```

也就是接下来的开发中，会传递Message对象的what值为0，接着就将目前时间呈现在一个变量名称为msg2的TextView对象中。以下假设有一个MyHandler类别对象已经建构为对象变量handler了。

自定义一个TimerTask类别的子类别：

```
private class MyTask1 extends TimerTask {
    @Override
    public void run( ) {
         handler.sendEmptyMessage(0);
    }
}
```

其工作任务内容就是发送Message而已。

也在建构出对象实体为task1之后，将该工作任务排进Timer对象中。

```
timer.schedule(task1, 0, 200);
```

设定一开始不延迟马上执行任务，之后每间隔 200个1/1000s，也就是0.2s的周期再度执行一次。

再做一个不同的TimerTask的工作内容，也同时排在相同的Timer对象的时间管理中。

```
private class MyTask2 extends TimerTask {
    @Override
    public void run( ) {
         handler.sendEmptyMessage((int)(Math.random( )*49+1));
    }
}
```

只比上一个复杂一点点，随机产生1～49的大乐透号码。

也稍微修改了MyHandler类别中handleMessage( )方法的程序内容。

```
@Override
public void handleMessage(Message msg) {
    if (msg.what == 0){
         Calendar now = Calendar.getInstance( );
         now.setTime(new Date(System.currentTimeMillis( )));
         msg2.setText(now.get(Calendar.MINUTE) + ":" + now.
           get(Calendar.SECOND));
```

```
    }else {
        msg1.setText("" + msg.what);
    }
}
```

并开发一个启动方法与一个停止方法。

```
private void startTask2( ){
    if (task2 == null){
        task2 = new MyTask2( );
        //timer.schedule(task2, 0, 1000);
        timer.scheduleAtFixedRate(task2, 0, 1000);
    }
}
private void endTask2( ){
    if (task2 != null){
        task2.cancel( );
        task2 = null;
    }
}
```

虽然是一样的工作任务，让用户可以进行启动与暂停，关键动作在于调用 TimerTask 对象的 cancel( ) 方法，可以使该对象结束在 Timer 中的时间周期。而如果希望一次全部在 Timer 中的所有任务都取消，则将会调用 Timer 对象的 cancel( ) 方法。

即使 App 在用户最后按下返回键之后，Timer 的运作仍然在后台中持续进行。如果希望应该一并取消，则可以在 Activity 中 Override 其 finish( ) 方法中处理。

```
@Override
public void finish( ) {
    timer.cancel( );
    timer = null;
    super.finish( );
}
```

## 5-3 同步任务AsyncTask

android.os.AsyncTask是android中提供一种方便、容易使用的用户线程（UI Thread），可以用来处理后台中执行程序，并可以将一或多个结果直接呈现在用户界面上，而完全不需要通过Handler的机制，算是一种相当方便的处理机制。

### ■ 5-3-1 使用观念

① 开发自定义类别继承AsyncTask。

② 指定其三个泛型参数类型，不可以是基本类型，若不指定，则必须使用Void。这三项参数在开发中指定，才能充分发挥出AsyncTask直接而不通过Handler机制来处理用户界面的优点。

③ Override其以下方法：

- doInBackground( )。
- onCancelled( )。
- onPostExecute( )。
- onPreExecute( )。
- onProgressUpdate( )。

### ■ 5-3-2 生命周期

先从其生命周期的表现来观察处理程序。因为在认识了解阶段，所以三个泛型参数类型皆设定为Void。

```java
private class MyTask extends AsyncTask<Void,Void,Void> {
    @Override
    protected Void doInBackground(Void...params) {
        Log.i("brad", "doInBackground");
        return null;
    }

    @Override
    protected void onCancelled( ) {
        super.onCancelled( );
        Log.i("brad", "onCancelled");
    }
```

```java
@Override
protected void onPostExecute(Void result) {
    super.onPostExecute(result);
    Log.i("brad", "onPostExecute");
}
@Override
protected void onPreExecute( ) {
    super.onPreExecute( );
    Log.i("brad", "onPreExecute");
}
@Override
protected void onProgressUpdate(Void... values) {
    super.onProgressUpdate(values[0]);
    Log.i("brad", "onProgressUpdate" + values[1]);
}
}
```

并撰写一段程序来进行测试，包含两个Button，分别作用为建构AsyncTask对象实体，并开始执行；另一个用来提早结束AsyncTask对象实体的生命周期。

```java
start = findViewById(R.id.start);
end = findViewById(R.id.end);
start.setOnClickListener(new OnClickListener( ) {
    @Override
    public void onClick(View v) {
        task = new MyTask( );
        task.execute("Brad", null, null);
    }
});
end.setOnClickListener(new OnClickListener( ) {
    @Override
    public void onClick(View v) {
        if (task != null && !task.isCancelled( )){
            task.cancel(true);
        }
    }
});
```

- start 与 end 为定义在 layout 中的 Button。
- task 为声明为 MyTask 的对象变量。

执行结果将会观察到如下画面。

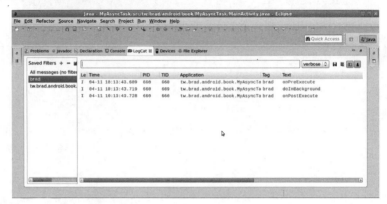

表示在基本程序中的顺序为：
- onPreExecute( )：执行前的准备工作。
- doInBackground( )：主要执行内容。
- onPostExecute( )：执行结束后的处理程序。

接着在 doInBackground( ) 中调用 publishProgress( )。

```
@Override
protected Void doInBackground(Void...params) {
    Log.i("brad", "doInBackground");
    publishProgress( );
    return null;
}
```

发现其执行 onProgressUpdate( ) 方法，这就是重点所在的地方，AsyncTask 在主要执行 doInBackground( ) 中可以随时调用 publishProgress( )，并传递要表现在用户界面的参数，而由 onProgressUpdate( ) 来负责更新用户界面。

而当调用 AsyncTask 对象实体的 cancel( ) 方法则会使其提早结束生命周期后，执行 onCancelled( ) 方法，通常用来辨别处理提早结束的程序使用。

### ■ 5-3-3 定义泛型参数

① 定义用来传递给 doInBackground( ) 方法的参数类型，通常会是在执行中会用到外部传递给线程的参数。

② 定义 doInBackground( ) 中调用 publishProgress( ) 时传递的参数类型，通常会是表现在用户界面的数据类型。

③ 定义最后结果的数据类型。

### 5-3-4 基本开发程序

开发程序只要观念正确，就会非常简单容易。

假设会传递多个String对象参数，并会传回String对象参数结果，中间过程传递多个Integer对象参数。所以：

```
private class MyTask extends AsyncTask<String, Integer, String>
```

接着定义 doInBackground( )：

```
protected String doInBackground(String... names)
```

接收参数类型与宣告第三个泛型类型是一样的，而传回类型与第三个泛型类型是一样的。

当执行完毕之后，定义 onPostExecute( )方法：

```
protected void onPostExecute(String result)
```

接收参数类型与声明第三个泛型类型是一样的。处理执行过程的显示用户界面：

```
protected void onProgressUpdate(Integer... values)
```

接收参数类型与声明第二个泛型类型是一样的。

### 5-3-5 程序架构

整体程序代码架构如下：

activity_main.xml

```xml
<LinearLayout xmlns:android="http://schemas.android.com/apk/res/android"
    android:layout_width="match_parent"
    android:layout_height="match_parent"
    android:orientation="vertical"
    >
    <Button
        android:id="@+id/start"
        android:layout_width="match_parent"
```

```xml
            android:layout_height="wrap_content"
            android:text="Start"
            />
    <Button
            android:id="@+id/end"
        android:layout_width="match_parent"
        android:layout_height="wrap_content"
        android:text="End"
        />
    <TextView
            android:id="@+id/msg"
        android:layout_width="match_parent"
        android:layout_height="wrap_content"
        />
</LinearLayout>
```

而主要程序如下：

MainActivity.java

```java
package tw.brad.android.book.MyAsyncTask;

import android.app.Activity;
import android.os.AsyncTask;
import android.os.Bundle;
import android.view.View;
import android.view.View.OnClickListener;
import android.widget.TextView;

public class MainActivity extends Activity {
    private View start, end;
    private TextView mesg;
    private MyTask task;

    @Override
    protected void onCreate(Bundle savedInstanceState) {
        super.onCreate(savedInstanceState);
        setContentView(R.layout.activity_main);
```

```java
        start = findViewById(R.id.start);
        end = findViewById(R.id.end);
        mesg = (TextView) findViewById(R.id.msg);
        start.setOnClickListener(new OnClickListener( ) {
            @Override
            public void onClick(View v) {
                task = new MyTask( );
                //传递多个参数给AsyncTask运行处理
                task.execute("Brad", "Vivian", "Daniel");
            }
        });

        end.setOnClickListener(new OnClickListener( ) {
            @Override
            public void onClick(View v) {
                // 提早结束AsyncTask
                if (task != null && !task.isCancelled( )) {
                    task.cancel(true);
                }
            }
        });
    }

    private class MyTask extends AsyncTask<String, Integer, String> {
        private String[] name;
        private boolean isOver;

        @Override
        protected String doInBackground(String... names) {
            name = names;
            for (int i = 0; i < 20; i++) {
                try {
                    // 传递三个参数处理用户界面
                    publishProgress(i, i + 100, i+200);
                    Thread.sleep(500);    // 暂停休眠

                    // 如果提早结束
                    if (isCancelled( )) {
```

```java
                            isOver = true;
                            break;
                    }
                } catch (InterruptedException e) {
                    e.printStackTrace( );
                }
            }
            // 提早结束传会null；否则正常下传回结果参数
            return isOver?null:"终于执行结束";
    }

    @Override
    protected void onCancelled( ) {
        super.onCancelled( );
        mesg.setText("提早结束");
    }

    @Override
    protected void onPostExecute(String result) {
        super.onPostExecute(result);
        // 将接收执行结束后的结果参数显示在用户界面
        mesg.setText("Result:" + result);
    }

    @Override
    protected void onPreExecute( ) {
        super.onPreExecute( );
    }

    @Override
    protected void onProgressUpdate(Integer... values) {
        super.onProgressUpdate( );
        int i = values[0];
        // 将执行过程的传递参数显示在用户界面
        mesg.setText(name[i%3] + " = " + values[i%3]);
    }
}
}
```

## 5-4 倒数定时器

字面上的意思似乎用来进行倒数计时使用，例如跨年倒数计时、码表计时等。其实还有许多场合常用到，例如在许多的游戏 App 中，当宝物或是道具出现是有时间性，出现 8s 后自动消失，并会在消失前 3s 以闪烁方式提醒用户，这样的状况下就非常简单地以 android.os.CountDownTimer 来处理。或是，某个关卡必须限定在 20s 内闯关，否则就算是失败，这样的情境也非常适用。如果以前面小节的线程来处理，相对地在开发上就显得复杂许多。

### ■ 5-4-1 开发模式

① 自定义类别继承 CountDownTimer。
② Override 其建构式。
③ Override 以下方法：
- onTick( ): 时间间隔的处理程序。
- onFinish( ): 时间到的处理程序。

结构式中所传递的两个参数为：
- long 类型的总运行时间，单位为 1/1000s。
- long 类型的间隔时间，单位为 1/1000s。

 而在 onTick( ) 和 onFinish( ) 中可以直接更新用户界面，不需要通过 Handle 的机制，就是最方便的地方。

### ■ 5-4-2 直接实作练习

① activity_main.xml

```
<LinearLayout xmlns:android="http://schemas.android.com/apk/res/android"
    android:layout_width="match_parent"
    android:layout_height="match_parent"
    android:orientation="vertical"
    >
    <Button
        android:id="@+id/start"
        android:layout_width="match_parent"
        android:layout_height="wrap_content"
```

```xml
            android:text="Start"
            />
        <Button
            android:id="@+id/cancel"
            android:layout_width="match_parent"
            android:layout_height="wrap_content"
            android:text="Cancel"
            />
        <TextView
            android:id="@+id/tv"
            android:layout_width="wrap_content"
            android:layout_height="wrap_content"
            android:text="@string/hello_world"
            />
</LinearLayout>
```

② MainActivity.java

```java
package tw.brad.android.book.MyCountDownTest;

import android.app.Activity;
import android.os.Bundle;
import android.os.CountDownTimer;
import android.view.View;
import android.view.View.OnClickListener;
import android.widget.TextView;

public class MainActivity extends Activity {
    private View start, cancel;
    private TextView tv;
    private MyCoundDownTask mytask;

    @Override
    protected void onCreate(Bundle savedInstanceState) {
        super.onCreate(savedInstanceState);
        setContentView(R.layout.activity_main);

        tv = (TextView)findViewById(R.id.tv);
        start = findViewById(R.id.start);
        cancel = findViewById(R.id.cancel);
```

```java
            start.setOnClickListener(new OnClickListener( ) {
                @Override
                public void onClick(View v) {
                    startCountDown( );
                }
            });
            cancel.setOnClickListener(new OnClickListener( ) {
                @Override
                public void onClick(View v) {
                    cancelCountDown( );
                }
            });
    }
    private void startCountDown( ){
        mytask = new MyCoundDownTask(20000, 1000);
        mytask.start( );
    }

    private void cancelCountDown( ){
        if (mytask != null){
            mytask.cancel( );
            tv.setText("提早结束");
        }
    }
    private class MyCoundDownTask extends CountDownTimer {
        public MyCoundDownTask(long millisInFuture, long countDownInterval) {
            super(millisInFuture, countDownInterval);
        }

        @Override
        public void onTick(long millisUntilFinished) {
            tv.setText("剩下秒数: " + millisUntilFinished/1000);
        }

        @Override
        public void onFinish( ) {
            tv.setText("时间到");
        }
    }
}
```

# MEMO

# 第6课　菜单与动作列处理

6-1　菜单 Menu

6-2　动作列 Action Bar

## 6-1 菜单Menu

### 6-1-1 Options menu选项菜单（硬件菜单键）

当Activity在执行状态下时，用户按下移动装置的菜单键，通常在屏幕下方将会出现菜单的功能。一般而言，用在较不常见的功能使用，例如辅助说明或是关于功能。官方建议在Android 3+以后的版本，将会以Action Bar取代之，该部分将会在下一小节中介绍使用。

（1）事先定义的菜单内容

通常会以XML格式来表现出菜单内容，而XML格式文件会被放在项目架构下res/目录下的menu/子目录中。目前的开发环境下，当建立出一个新的Android项目后，都已经自动产生预设的文件，放在res/menu/main.xml，其内容如下：

```xml
<menu xmlns:android="http://schemas.android.com/apk/res/android" >
    <item
        android:id="@+id/action_settings"
        android:orderInCategory="100"
        android:showAsAction="never"
        android:title="@string/action_settings"/>
</menu>
```

而在主要的Activity中也已经写好该段处理方法，如下：

```java
@Override
public boolean onCreateOptionsMenu(Menu menu) {
    //Inflate the menu;this adds items to the action bar if it is present.
    getMenuInflater( ).inflate(R.menu.main, menu);
    return true;
}
```

直接执行后，当按下硬件菜单键后出现如下图所示。

假设想要设计出一个菜单内容为：

- 设定。
- 辅助。
- 关于。

则修改menu.xml基本如下：

```xml
<menu xmlns:android="http://schemas.android.com/apk/res/android" >
    <item
        android:id="@+id/action_settings"
        android:orderInCategory="200"
        android:showAsAction="never"
        android:title="Setting"
        />
    <item
        android:id="@+id/action_help"
        android:orderInCategory="400"
        android:showAsAction="never"
        android:title="Help"
        />
        android:id="@+id/action_about"
        android:orderInCategory="100"
        android:showAsAction="never"
        android:title="About"
    android:icon="@drawable/ic_launcher"
        />
</menu>
```

- 多加了两个<item>。
- 调整了orderInCategory的值，数值表示出现的优先级，小先大后。
- 第三个<item>，增设icon，指定为res/drawable/下的ic_launcher的影像文件。

呈现如下内容：

（2）侦测用户操作

回到程序开发工作区中，在类别中按下鼠标右键出现菜单，点选 Source 后再点选 Override/Implement Methods 出现如下画面：

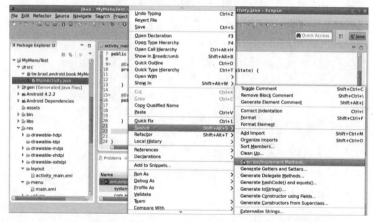

接着出现以下对话框画面，勾选 onOptionsItemSeleted( )。

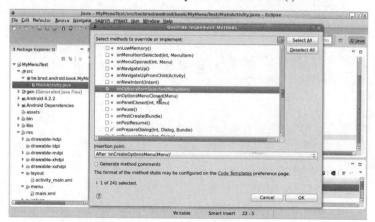

程序中自动产生以下程序代码：

```
@Override
public boolean onOptionsItemSelected(MenuItem item) {
    // TODO Auto-generated method stub
    return super.onOptionsItemSelected(item);
}
```

这段方法的开发撰写，是针对当用户点选了菜单中的特定项目之后所执行的方法。重点在于将会传递 MenuItem 对象的参数进来，该 MenuItem 对象即为用户所按下的菜单项目。接着通过调用 MenuItem 对象实体的 getItemId( ) 方法传回一个整数值，代表其在 res/menu/ 下的配置资源整数值，如下进行判断：

```
@Override
public boolean onOptionsItemSelected(MenuItem item) {
    switch (item.getItemId( )) {
    case R.id.action_about:
            Toast.makeText(this, "选择了about", Toast.LENGTH_SHORT).show( );
            break;
    case R.id.action_help:
            Toast.makeText(this, "选择了help", Toast.LENGTH_SHORT).show( );
            break;
    case R.id.action_settings:
            Toast.makeText(this, "选择了setting", Toast.LENGTH_SHORT).show( );
            break;
    }
    return super.onOptionsItemSelected(item);
}
```

## ■ 6-1-2　Context menu 内容菜单

内容菜单通常始于 ListView 之类的 View 一起运作。例如当在一个 ListView 的多个选项中，用户针对特定的选项进行长按，将会出现一个浮出菜单，引导用户进行该选项的进一步处理选择，而每个选项都有一样的浮出菜单。假设 ListView 中呈现出的是文件列表，用户长按特定的文件项目，就会浮现出一个浮出菜单，引导用户进行文件复制、更名或是删除。

（1）初始环境

以下以范例实际进行开发，假设在一个 ListActivity 中呈现一个 ListView，项目结构如下处理：

res/layout/activity_main.xml 负责处理 ListView 的单一选项版面。

```xml
<RelativeLayout xmlns:android="http://schemas.android.com/apk/res/android"
    android:layout_width="match_parent"
    android:layout_height="match_parent"
    >
    <TextView
        android:id="@+id/filename"
        android:layout_width="wrap_content"
        android:layout_height="wrap_content"
        android:textSize="24sp"
        android:textStyle="bold"
        />
</RelativeLayout>
```

只有一个文字内容，其 id 为 filename。

MainActivity 继承 ListActivity 类别，假设数据已经处理完毕（因为本单元并非介绍 ListView，所以数据来源是从程序中产生，实际状况应该是从文件系统中，捞取数据放进菜单中）。

```java
public class MainActivity extends ListActivity
    private ListView lview;
    private SimpleAdapter adapter;
    private String[] from = {"filename"};
    private int[] to = {R.id.filename};
    private ArrayList<HashMap<String,String>> data;

    @Override
    protected void onCreate(Bundle savedInstanceState) {
        super.onCreate(savedInstanceState);

        data = new ArrayList<HashMap<String,String>>( );

        HashMap<String,String> d0 = new HashMap<String,String>( );
        d0.put(from[0], "file0.txt");
        data.add(d0);
        HashMap<String,String> d1 = new HashMap<String,String>( );
        d1.put(from[0], "file1.txt");
```

```
            data.add(d1);
            HashMap<String,String> d2 = new HashMap<String,String>( );
            d2.put(from[0], "file2.txt");
            data.add(d2);
            HashMap<String,String> d3 = new HashMap<String,String>( );
            d3.put(from[0], "file3.txt");
            data.add(d3);
            adapter = new SimpleAdapter(this, data, R.layout.activity_
            main, from, to);
            lview = getListView( );
            lview.setAdapter(adapter);
        }
    }
```

因此呈现如下图的初始状态：

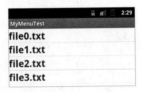

（2）建立 ContextMenu 菜单

开始处理 ContextMenu 菜单，在 res/menu/ 子目录下，新建一个菜单名为 cmenu.xml。

```
<?xml version="1.0" encoding="utf-8"?>
<menu xmlns:android="http://schemas.android.com/apk/res/android" >
    <item
        android:id="@+id/item1"
        android:title="编辑"
        />
    <item
        android:id="@+id/item2"
        android:title="删除"
        />
    <item
        android:id="@+id/item3"
        android:title="列表"
        />
```

```xml
    <item
        android:id="@+id/item4"
        android:title="测试"
        />
</menu>
```

- Id用来设定资源内容。
- title用来设定显示内容。

回到程序中，针对 ListView 或是 GridView 中进行注册 ContextMenu，也就是调用 registerForContextMenu( ) 方法，例如：

```
registerForContextMenu(lview);
```

在 Activity 类别中，Override 其 onCreateContextMenu( ) 方法：

```
@Override
public void onCreateContextMenu(ContextMenu menu, View v, ContextMenuInfo menuInfo) {
    super.onCreateContextMenu(menu, v, menuInfo);
    getMenuInflater( ).inflate(R.menu.cmenu, menu);
}
```

开发到这边就已经可以先执行查看状况：

（3）侦测用户的操作行为

处理手法与 OptionMenu 的处理手法类似即可。

```
@Override
public boolean onContextItemSelected(MenuItem item) {
```

```
    switch (item.getItemId( )){
        case R.id.item1:
            Toast.makeText(this, "编辑", Toast.LENGTH_SHORT).show( );
            break;
        case R.id.item2:
            Toast.makeText(this, "删除", Toast.LENGTH_SHORT).show( );
            break;
        case R.id.item3:
            Toast.makeText(this, "列表", Toast.LENGTH_SHORT).show( );
            break;
        case R.id.item4:
            Toast.makeText(this, "测试", Toast.LENGTH_SHORT).show( );
            break;
    }
    return super.onContextItemSelected(item);
}
```

## ■ 6-1-3　Popup menu弹出式菜单

弹出式菜单用在View组件上面，用户按下该View组件之后，就会出现一个弹出的菜单。不过要特别注意的是，该项功能适用于API Level 11+，也就是Android 3.0+之后的版本。

先看到其基本的处理结果如下图所示：

处理程序：
- 在res/menu/子目录下建立菜单XML菜单内容文件。
- 设定特定的View的按下事件。
- 建构PopupMenu。
- 指定res/menu/下的XML菜单内容文件。
- 设定特定选项的事件。

在res/menu/子目录下建立菜单XML菜单内容文件：res/menu/pmenu.xml。

```
<?xml version="1.0" encoding="utf-8"?>
```

```xml
<menu xmlns:android="http://schemas.android.com/apk/res/android" >
    <item android:id="@+id/item1" android:title="Test 1"></item>
    <item android:id="@+id/item2" android:title="Test 2"></item>
</menu>
```

主要版面 activity_main.xml,指定 Button 的按下 onClick 为调用 show PopupMenu 方法。

```xml
<LinearLayout xmlns:android="http://schemas.android.com/apk/res/android"
    android:layout_width="match_parent"
    android:layout_height="match_parent"
    android:orientation="vertical"
    >

    <Button
        android:id="@+id/click1"
        android:layout_width="wrap_content"
        android:layout_height="wrap_content"
        android:text="Click 1"
        android:onClick="showPopupMenu"
        />

    <TextView
        android:id="@+id/info"
        android:layout_width="wrap_content"
        android:layout_height="wrap_content"
        android:text="@string/hello_world"
        />

</LinearLayout>
```

回到主要程序中:

```java
public void showPopupMenu(View v){
    PopupMenu popup = new PopupMenu(this, v);
    MenuInflater inflater = popup.getMenuInflater( );
    inflater.inflate(R.menu.pmenu, popup.getMenu( ));
    popup.show( );
}
```

再来进行按下侦测，处理手法与之前OptionMenu类似：

```
public void showPopupMenu(View v){
    PopupMenu popup = new PopupMenu(this, v);
    MenuInflater inflater = popup.getMenuInflater( );
    inflater.inflate(R.menu.pmenu, popup.getMenu( ));

    popup.setOnMenuItemClickListener(new PopupMenu.OnMenuItemClickListener( ) {
            @Override
            public boolean onMenuItemClick(MenuItem item) {
                    switch (item.getItemId( )) {
                            case R.id.item1:
                                    Toast.makeText(MainActivity.this, "编辑",
Toast.LENGTH_SHORT).show( );
                                    break;
                            case R.id.item2:
                                    Toast.makeText(MainActivity.this, "删除",
Toast.LENGTH_SHORT).show( );
                                    break;
                    }
                    return false;
            }
    });
    popup.show( );
}
```

## 6-2 动作列Action Bar

在一般建立项目精灵对话框之后，其实已经含进了在上一小节中所介绍的OptionMenu，只需要再略作些修改就是Action Bar。

先将res/menu/子目录下的main.xml进行修改如下：

```
<menu xmlns:android="http://schemas.android.com/apk/res/android" >

    <item
        android:id="@+id/action_share"
```

```xml
        android:icon="@drawable/share"
        android:showAsAction="ifRoom|withText"
        android:title="分享"
        />
<item
        android:id="@+id/action_gift"
        android:icon="@drawable/gift"
        android:showAsAction="ifRoom|withText"
        android:title="礼物"
        />
<item
        android:id="@+id/action_delete"
        android:icon="@drawable/delete"
        android:showAsAction="ifRoom|withText"
        android:title="删除"
        />
<item
        android:id="@+id/action_eraser"
        android:icon="@drawable/eraser"
        android:showAsAction="ifRoom|collapseActionView"
        android:title="清除"
        />
<item
        android:id="@+id/action_save"
        android:icon="@drawable/save"
        android:showAsAction="ifRoom|collapseActionView"
        android:title="存档"
        />
<item
        android:id="@+id/action_open"
        android:icon="@drawable/open"
        android:showAsAction="collapseActionView|ifRoom"
        android:title="开启"
        />
<item
        android:id="@+id/action_help"
```

```
        android:icon="@drawable/help"
        android:showAsAction="collapseActionView"
        android:title="辅助"
        />
</menu>
```

差别在于showAsAction的设定值。
- never：不显示在Action Bar。
- always：尽量显示在Action Bar，空间不够仍然要用硬件菜单键。
- ifRoom：如果Action Bar的空间还够就显示。
- with Text：也显示文字内容（通常在平板计算机上有作用）。
- collapseActionView：折叠显示，折叠展开后显示文字内容。

在API Level 3以前的装置，仍然视为OptionMenu处理，画面如下：

在平板电脑的状况如下图：

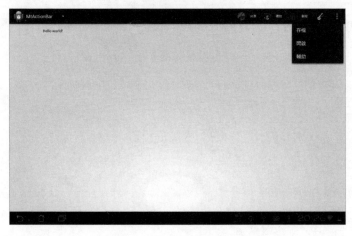

在 API Level 3+ 的手机的状况如下图：

而用户选取项目的判断方式也与 OptionMenu 相同。

```
@Override
public boolean onOptionsItemSelected(MenuItem item) {
    if (item.getItemId( ) == R.id.action_gift){
        Log.i("brad", "Gift");
    }else {
        Log.i("brad", "Other");
    }
    return super.onOptionsItemSelected(item);
}
```

如果在 menu 中，要加上搜寻的功能，可以在 res/menu/main.xml 中增加一个 item 如下：

```
<item
    android:id="@+id/action_search"
    android:showAsAction="ifRoom|withText"
    android:actionViewClass="android.widget.SearchView"
    android:title="搜寻"
    />
```

就会有以下的效果呈现：

当用户按下搜寻图标后，将会展开输入搜寻文字。

此时应该在程序中针对 SearchView 进行设定，先通过 menu 对象实体调用 findItem( ) 方法找出 id 为 action_search 的 item，再调用 getActionView( ) 方法后将返回值强制转型回 SearchView 对象实体。

```
SearchView searchView = (SearchView)menu.findItem(R.id.action_search).
getActionView( );
```

设定带有送出查询按钮。

```
searchView.setSubmitButtonEnabled(true);
```

设定当按下送出按钮后，接收用户输入的字符串。

```
searchView.setOnQueryTextListener(new OnQueryTextListener( ) {
    @Override
    public boolean onQueryTextSubmit(String query) {
        Log.i("brad", "查询字符串: " + query);
        return false;
    }
    @Override
    public boolean onQueryTextChange(String newText) {
        return false;
    }
});
```

再来增加设定一个共享提供者，可以将数据由用户决定给特定的应用程序来处理。先在 res/menu/main.xml 中增设一个 item。

```
<item android:id="@+id/menu_share"
    android:title="Share"
    android:showAsAction="ifRoom"
    android:actionProviderClass="android.widget.ShareActionProvider" />
```

回到程序的处理手法与 SearchView 类似（API Level 14+ 适用）。

```
// API Level 14+
ShareActionProvider sap = (ShareActionProvider)menu.findItem(R.id.menu_share)
.getActionProvider( );
sap.setShareIntent(getDefaultShareIntent( ));
```

而 getDefaultShareIntent( ) 方法是自定义设定共享提供者。

```
private Intent getDefaultShareIntent( ){
    Intent intent = new Intent(Intent.ACTION_SEND);
    intent.setType("text/plain");
    intent.putExtra(Intent.EXTRA_SUBJECT, "SUBJECT");
    intent.putExtra(Intent.EXTRA_TEXT,"Extra Text");
    return intent;
}
```

就会依照用户在移动装置上的 App 安装而定，例如下图。

不要有 Action Bar：

```
<activity android:theme="@android:style/Theme.Holo.NoActionBar">
```

或是在程序中隐藏：

```
ActionBar actionBar = getActionBar( );
actionBar.hide( );
```

# MEMO

# 07 Chapter

## 第7课　自定义View与SurfaceView

7-1　自定义View：继承View

7-2　自定义View与触控手势事件处理

7-3　自定义SurfaceView：继承SurfaceView

7-4　以自定义View来实现手写签名App范例实作

## 7-1　自定义View：继承View

在Android SDK中的android.view.View是用来呈现给用户的视觉界面组件，而在API中虽然提供非常丰富的View组件，包含TextView、Button等，但是开发者经常可能会发现想要的外观，操作互动等许多方面是现有API无法可以替代使用的状况，此时可以对View组件进行继承及自定义内容。

首先开发自定义类别extends android.view.View：

```
package tw.brad.book.myviewtest;

import android.content.Context;
import android.util.AttributeSet;
import android.view.View;

public class PaintView extends View {

}
```

其建构有三种，择其一进行Override，如果是采用只有传递一个Context类型的参数建构式：

```
public PaintView(Context context) {
    super(context);
}
```

则该自定义View类别无法在版面配置的XML文件中进行配置，只能在程序运行时间加进ViewGroup之中；如果想要设计出可以在版面配置的XML中进行规划，也可以在程序运行时间处理，则至少要Override的建构式是含有AttributeSet类型的参数。

```
public PaintView(Context context, AttributeSet attrs) {
    super(context, attrs);
}
```

或是

```
public PaintView(Context context, AttributeSet attrs, int defStyle) {
    super(context, attrs, defStyle);
}
```

此时可以先来进行版面配置处理。假设想要规划设计以横向为主的画面，左边有一排Button，用来进行触发用户的操作，而右边一整块区域用来放置自定义View。如下图所示的初始状态：

自定义View为PaintView.java。

```java
package tw.brad.book.myviewtest;

import android.content.Context;
import android.util.AttributeSet;
import android.view.View;

public class PaintView extends View {

    public PaintView(Context context, AttributeSet attrs) {
        super(context, attrs);
    }
}
```

版面配置activity_main.xml。

```xml
<LinearLayout xmlns:android="http://schemas.android.com/apk/res/android"
    android:layout_width="match_parent"
    android:layout_height="match_parent"
    android:orientation="horizontal" >

    <LinearLayout
        android:layout_width="match_parent"
        android:layout_height="match_parent"
        android:layout_weight="4"
        android:background="#1100ff00"
        android:orientation="vertical" >

        <Button
            android:id="@+id/clear"
```

```xml
            android:layout_width="match_parent"
            android:layout_height="wrap_content"
            android:layout_weight="1"
            android:text="Clear" />

        <Button
            android:id="@+id/undo"
            android:layout_width="match_parent"
            android:layout_height="wrap_content"
            android:layout_weight="1"
            android:text="Undo" />

        <Button
            android:id="@+id/redo"
            android:layout_width="match_parent"
            android:layout_height="wrap_content"
            android:layout_weight="1"
            android:text="Redo" />

        <Button
            android:id="@+id/chcolor"
            android:layout_width="match_parent"
            android:layout_height="wrap_content"
            android:layout_weight="1"
            android:text="Color" />

        <Button
            android:id="@+id/chsize"
            android:layout_width="match_parent"
            android:layout_height="wrap_content"
            android:layout_weight="1"
            android:text="Size" />

    </LinearLayout>

    <LinearLayout
        android:layout_width="match_parent"
        android:layout_height="match_parent"
        android:layout_weight="1"
        android:orientation="vertical" >
```

```xml
        <tw.brad.book.myviewtest.PaintView
            android:id="@+id/pview"
            android:layout_width="match_parent"
            android:layout_height="match_parent"
            android:background="#44ffff00" />
    </LinearLayout>
</LinearLayout>
```

为了避免因为移动装置的手持方式改变,而切换到直式显示,可以先在AndroidManifest.xml中进行设定。

```xml
<?xml version="1.0" encoding="utf-8"?>
<manifest xmlns:android="http://schemas.android.com/apk/res/android"
    package="tw.brad.book.myviewtest"
    android:versionCode="1"
    android:versionName="1.0" >

    <uses-sdk
        android:minSdkVersion="8"
        android:targetSdkVersion="17" />

    <application
        android:allowBackup="true"
        android:icon="@drawable/ic_launcher"
        android:label="@string/app_name"
        android:theme="@style/AppTheme" >
        <activity
          android:name="tw.brad.book.myviewtest.MainActivity"
          android:label="@string/app_name" android:screenOrientation="landscape">
            <intent-filter>
                <action android:name="android.intent.action.MAIN" />
                <category android:name="android.intent.category.LAUNCHER" />
            </intent-filter>
        </activity>
    </application>

</manifest>
```

接下来的开发重点会先放在PaintView的处理。

先来了解如何取得目前的自定义View的宽高值。对于View而言，本来就有getWidth( )与getHeight( )两个方法可以使用，就先在建构式中抓取这两个值。

```java
public PaintView(Context context, AttributeSet attrs) {
    super(context, attrs);

    int viewW = getWidth( );
    int viewH = getHeight( );
    Log.i("brad", viewW + " x " + viewH);
}
```

执行之后，取得的两个值皆为0。

而View的显示程序主要是在onDraw( )方法中进行，因此来Override该方法，并将取得宽高值在该方法中进行看看。

```java
@Override
protected void onDraw(Canvas canvas) {
    int viewW = getWidth( );
    int viewH = getHeight( );
    Log.i("brad", viewW + " x " + viewH);
}
```

终于得到了正确的值，也就是说在其呈现内容时才开始有了具体的宽高。如果自定义View每次进行呈现内容变动，还要重复抓取宽高值，在逻辑上应该不是如此。所以可以用一个boolean变量来判断是否是首次执行，如果是就当时抓取宽高值，若否就不再执行抓取宽高值。

结构更改如下：

```java
package tw.brad.book.myviewtest;

import android.content.Context;
import android.graphics.Canvas;
import android.util.AttributeSet;
import android.util.Log;
import android.view.View;

public class PaintView extends View {
    private boolean isInited;
    private int viewW, viewH;
```

```java
    public PaintView(Context context, AttributeSet attrs) {
        super(context, attrs);
    }
    private void init( ){
        viewW = getWidth( );
        viewH = getHeight( );
        isInited = true;
    }
    @Override
    protected void onDraw(Canvas canvas) {
        if (!isInited) init( );
    }
}
```

通过传递开始在onDraw( )方法中进行绘制,原理就是利用传递进来的Canvas对象进行绘制。

画个圆/文字/影像资源:

```java
@Override
protected void onDraw(Canvas canvas) {
    if (!isInited) init( );
    // 画出一个几何圆
    Paint paintCircle = new Paint( );
    paintCircle.setColor(Color.GREEN);
    canvas.drawCircle(viewW/2, viewH/2, 40, paintCircle);
    // 画出文字内容
    Paint paintText = new Paint( );
    paintText.setColor(Color.BLACK);
    paintText.setTextSize(36);
    paintText.setTextScaleX(1.5f);
    canvas.drawText("Brad Big Company", 40, 100, paintText);
    //  画出影像图文件
    Bitmap shield = BitmapFactory.decodeResource(getResources( ), R.drawable.shield);
    canvas.drawBitmap(shield, viewW/2+ 50, viewH/2, null);
}
```

结果如下：

这样的写法也不太好，因为在 onDraw( ) 方法中尽量能简化到只做绘制的工作，而其他的对象建构及设定等程序，可以事先定义好即可。稍微更改结构如下：

```java
package tw.brad.book.myviewtest;

import android.content.Context;
import android.graphics.Bitmap;
import android.graphics.BitmapFactory;
import android.graphics.Canvas;
import android.graphics.Color;
import android.graphics.Paint;
import android.util.AttributeSet;
import android.view.View;

public class PaintView extends View {
    private boolean isInited;
    private int viewW, viewH;
    private Paint paintCircle, paintText;
    private Bitmap shield;
    public PaintView(Context context, AttributeSet attrs) {
        super(context, attrs);
    }

    private void init( ) {
        viewW = getWidth( );
        viewH = getHeight( );

        paintCircle = new Paint( );
        paintCircle.setColor(Color.GREEN);

        paintText = new Paint( );
        paintText.setColor(Color.BLACK);
```

```
            paintText.setTextSize(36);
            paintText.setTextScaleX(1.5f);
    shield = BitmapFactory.decodeResource(getResources( ), R.drawable.shield);
            isInited = true;
    }

    @Override
    protected void onDraw(Canvas canvas) {
            if (!isInited) init( );
        // 画出一个几何圆
        canvas.drawCircle(viewW / 2, viewH / 2, 40, paintCircle);
        // 画出文字内容
        canvas.drawText("Brad Big Company", 40, 100, paintText);
        // 画出影像图文件
        canvas.drawBitmap(shield, viewW / 2 + 50, viewH / 2, null);
    }
}
```

再来练习通过线程来不断重新绘制Canvas对象，而能够达成视觉上动态的效果。不断重新绘制就是调用postInvalidate( )方法，就会再度执行onDraw( )方法。周期任务用来改变其显示位置，以上例中的圆而言，就是改变其圆心位置。

```
package tw.brad.book.myviewtest;
import java.util.Timer;
import java.util.TimerTask;

import android.content.Context;
import android.graphics.Bitmap;
import android.graphics.BitmapFactory;
import android.graphics.Canvas;
import android.graphics.Color;
import android.graphics.Paint;
import android.util.AttributeSet;
import android.view.View;

public class PaintView extends View {
    private boolean isInited;
```

```java
        private int viewW, viewH, cx, cy;
        private Paint paintCircle, paintText;
        private Bitmap shield;
        private Timer timer;
        private MyTask task;

    public PaintView(Context context, AttributeSet attrs) {
            super(context, attrs);
            timer = new Timer( );
            task = new MyTask( );
    }

    private void init( ) {
            viewW = getWidth( );
            viewH = getHeight( );

            paintCircle = new Paint( );
            paintCircle.setColor(Color.GREEN);

            paintText = new Paint( );
            paintText.setColor(Color.BLACK);
            paintText.setTextSize(36);
            paintText.setTextScaleX(1.5f);

            shield = BitmapFactory.decodeResource(getResources( ),
R.drawable.shield);
            cx = cy = 50;

            timer.scheduleAtFixedRate(task, 0, 80);
            isInited = true;
    }

    @Override
    protected void onDraw(Canvas canvas) {
            if (!isInited) init( );

            // 画出一个几何圆
            canvas.drawCircle(cx, cy, 40, paintCircle);

            // 画出文字内容
            canvas.drawText("Brad Big Company", 40, 100, paintText);
```

```
            // 画出影像图文件
            canvas.drawBitmap(shield, viewW / 2 + 50, viewH / 2, null);
    }
    private class MyTask extends TimerTask {
        private int i;
        @Override
        public void run( ) {
            cx += 4;
            postInvalidate( );
            if (i++>100) cancel( );
        }
    }
}
```

重点整理如下：

• postInvalidate( )用来整个重新绘制全部呈现内容。

• 默认上绘制原理依照调用canvas.drawXxx( )的优先级，先画者位于较底层。例如背景内容先绘制，游戏主角通常最后绘制。

• 线程以内部类别方式处理较为容易开发，因为可以直接存取外部类别的成员。

## 7-2 自定义View与触控手势事件处理

### ■ 7-2-1 一般触控事件侦测处理

再来针对用户的触发事件处理，最常见的就是onTouchEvent( )。所以就开始进行Override其onTouchEvent( )方法：

```
@Override
public boolean onTouchEvent(MotionEvent event) {
    return super.onTouchEvent(event);
}
```

重点在于该方法被触发后传递进来的MotionEvent对象实体，包含了用户所触摸屏幕的像素坐标位置。

假设当用户触摸自定义View的位置时，将绘制的圆心移至此处。

```
@Override
public boolean onTouchEvent(MotionEvent event) {
    cx = (int)event.getX( );
    cy = (int)event.getY( );
    postInvalidate( );
    return super.onTouchEvent(event);
}
```

但是当用户摸着屏幕没有离开，直接滑动游走时，则没有任何反应发生。那是因为其return值super.onTouchEvent(event)为false，表示这次触摸行为不再重复触发。若将其return值设定为true，则将会不断地触发该事件，轻松地使圆跟着手指头移动。

### ■ 7-2-2 手势侦测事件处理

而另外还有一种常见的情境，就是想要设计出用户在自定义View对象实体上向上下左右划过，判断为方向控制处理。此时就需要利用到android.view.GestureDetector对象实体进行进一步的侦测。

先进行声明对象实体变量：

```
private GestureDetector gd;
```

开发一个自定义类别来实作GestureDetector.OnGestureListener界面，不过这样处理略为麻烦，如果只是要Override其方法而已，那就直接继承android.view.GestureDetector.SimpleOnGestureListener类别即可（光看名称就觉得简单容易）。

```
private class MyGDListener extends SimpleOnGestureListener {
    @Override
    public boolean onDown(MotionEvent e) {
        return true;
    }
    @Override
    public boolean onFling(MotionEvent e1, MotionEvent e2, float velocityX,
            float velocityY) {
        if (Math.abs(velocityX) > Math.abs(velocityY)){
            if (velocityX > 10){
                cx += 10;
            }else if (velocityX < -10){
```

```
                    cx -= 10;
                }
            }else{
                if (velocityY > 10){
                    cy += 10;
                }else if(velocityY < -10){
                    cy -= 10;
                }
            }
            postInvalidate( );
            return super.onFling(e1, e2, velocityX, velocityY);
        }
    }
}
```

重点说明：

• 主要目的在于Override其onFling( )方法。

• 但是会先被onDown( )方法侦测到，而其预设返回boolean值是表示不再往下侦测，若设定返回值为true，才会往下侦测动作。

• onFling( )中的第三个参数表示坐标X轴方向移动速度，单位为每秒移动像素，向右为正值，向左为负值；第四个参数表示坐标Y轴方向移动速度，单位为每秒移动像素，向下为正值，向上为负值而该手势侦测事件目前为止，完全没有触发点。所以，需要刚刚的onTouchEvent( )方法来启动触发事件，因此，将其return值改为gd.onTouch Event(event)即可。

整个架构如下：

```
package tw.brad.book.myviewtest;
import java.util.Timer;
import java.util.TimerTask;

import android.content.Context;
import android.graphics.Bitmap;
import android.graphics.BitmapFactory;
import android.graphics.Canvas;
import android.graphics.Color;
import android.graphics.Paint;
import android.util.AttributeSet;
import android.view.GestureDetector;
```

```java
import android.view.GestureDetector.SimpleOnGestureListener;
import android.view.MotionEvent;
import android.view.View;
public class PaintView extends View {
    private boolean isInited;
    private int viewW, viewH, cx, cy;
    private Paint paintCircle, paintText;
    private Bitmap shield;
    private Timer timer;
    private MyTask task;
    private GestureDetector gd;
    public PaintView(Context context, AttributeSet attrs) {
        super(context, attrs);
        timer = new Timer( );
        task = new MyTask( );
        gd = new GestureDetector(context, new MyGDListener( ));
    }
    private void init( ) {
        viewW = getWidth( );
        viewH = getHeight( );
        paintCircle = new Paint( );
        paintCircle.setColor(Color.GREEN);
        paintText = new Paint( );
        paintText.setColor(Color.BLACK);
        paintText.setTextSize(36);
        paintText.setTextScaleX(1.5f);
        shield = BitmapFactory
                .decodeResource(getResources( ), R.drawable.shield);
        cx = cy = 50;
        timer.scheduleAtFixedRate(task, 0, 80);
        isInited = true;
    }
    @Override
```

```java
    protected void onDraw(Canvas canvas) {
        if (!isInited)
            init( );
        // 画出一个几何圆
        canvas.drawCircle(cx, cy, 40, paintCircle);
        // 画出文字内容
        canvas.drawText("Brad Big Company", 40, 100, paintText);
        // 画出影像图文件
        canvas.drawBitmap(shield, viewW / 2 + 50, viewH / 2, null);
    }

    @Override
    public boolean onTouchEvent(MotionEvent event) {
        return gd.onTouchEvent(event);
    }
}
private class MyGDListener extends SimpleOnGestureListener {
    @Override
    public boolean onDown(MotionEvent e) {
        return true;
    }

    @Override
    public boolean onFling(MotionEvent e1, MotionEvent e2, float velocityX,
            float velocityY) {
        if (Math.abs(velocityX) > Math.abs(velocityY)){
            if (velocityX > 10){
                cx += 10;
            }else if (velocityX < -10){
                cx -= 10;
            }
        }else{
            if (velocityY > 10){
                cy += 10;
            }else if(velocityY < -10){
                cy -= 10;
            }
        }
```

```
                postInvalidate( );
                return super.onFling(e1, e2, velocityX, velocityY);
        }
}

    private class MyTask extends TimerTask {
        private int i;
        @Override
        public void run( ) {
            cx += 4;
            postInvalidate( );
            if (i++ > 100)
                cancel( );
        }
    }
}
```

## 7-3 自定义SurfaceView：继承SurfaceView

上两小节中介绍了自定义View的处理方式，其产生动态效果来自于周期更新绘制整个呈现的内容。对于简单的处理上来说，算是相当容易开发。但是当要呈现的内容较为复杂时，可能会有多个动态对象同时进行移动，则非常容易发生内存不足，或是效能不佳。

而本小节所要介绍的android.view.SurfaceView则可以针对单次要变动更新的区域进行锁定，才开始针对该锁定区域进行更新绘制，最后再进行解锁。因为只需要更新变动需要变动的区域，所以在处理效能上较佳。

开发的架构与自定义View非常类似，此小节就不以版面配置来处理。所以直接继承。

```
package tw.brad.book.mysvtest1;
import android.content.Context;
import android.view.SurfaceView;
public class MySV extends SurfaceView {
    public MySV(Context context) {
        super(context);
    }
}
```

Surface 呈现画面的控制是由类别 SurfaceHolder 对象实体来进行处理的，可以由调用 getHolder( ) 取得。

```java
public class MySV extends SurfaceView {
    private SurfaceHolder holder;
    public MySV(Context context) {
        super(context);
        holder = getHolder( );
    }
}
```

再由 SurfaceHolder 对象实体调用 addCallback( ) 方法，指定实作 Surface Holder.Callback 界面的对象实体。下例中直接由自定义 SurfaceView 来实作即可：

```java
public class MySV extends SurfaceView implements Callback {
    private SurfaceHolder holder;
    public MySV(Context context) {
        super(context);
        holder = getHolder( );
        holder.addCallback(this);
    }
    @Override
    public void surfaceChanged(SurfaceHolder holder, int format, int width,
        int height) {
        // 呈现画面变动
    }

    @Override
    public void surfaceCreated(SurfaceHolder holder) {
        // 建立呈现的画面
    }

    @Override
    public void surfaceDestroyed(SurfaceHolder holder) {
        // 结束呈现的画面
    }
}
```

其处理模式绘制画面方式，是通过 SurfaceHolder 对象实体调用 lockCanvas( ) 方法传回被锁定冻结的Canvas对象实体，再以Canvas对象实体进行绘制，最后再由SurfaceHolder对象实体调用unlockCanvasAndPost( )方法，传入已经绘制完成的Canvas对象实体为参数即可。

```
Canvas canvas = holder.lockCanvas(null);
canvas.drawBitmap(bg, 0, 0, null);
canvas.drawBitmap(shield, 350, 350, null);
canvas.drawCircle(x, y, 50, paint);
holder.unlockCanvasAndPost(canvas);
```

重点是当lockCanvas传入参数为null时，将回传整个绘制区域。

而通常首次绘制的时机点是在surfaceCreated( )方法的调用。

下面范例中以一个线程来重复绘制在特定区域：

```java
package tw.brad.book.mysvtest1;

import android.content.Context;
import android.graphics.Bitmap;
import android.graphics.BitmapFactory;
import android.graphics.Canvas;
import android.graphics.Color;
import android.graphics.Paint;
import android.graphics.Rect;
import android.view.SurfaceHolder;
import android.view.SurfaceHolder.Callback;
import android.view.SurfaceView;

public class MySurfaceView extends SurfaceView implements Callback {
    private SurfaceHolder holder;
    private Paint paint;
    private int x, y, dy;
    private boolean isInit;
    private Bitmap bg, shield;

    public MySurfaceView(Context context) {
        super(context);

        holder = getHolder( );
        holder.addCallback(this);

        paint = new Paint( );
        paint.setColor(Color.YELLOW);
```

```
            x = y = 100;
            dy = 12;
            bg = BitmapFactory.decodeResource(getResources( ), R.drawable.bg0);
            shield = BitmapFactory
                    .decodeResource(getResources( ), R.drawable.shield);
}
void drawCanvas( ) {
        Canvas canvas;
        if (!isInit) {
                canvas = holder.lockCanvas(null);
                canvas.drawBitmap(bg, 0, 0, null);
                canvas.drawBitmap(shield, 350, 350, null);
                isInit = true;
        } else {
                canvas = holder.lockCanvas(new Rect(x - 50, y - 50 -
                                        dy, x + 50, y + 50));
                canvas.drawBitmap(bg, 0, 0, null);
        }
        canvas.drawCircle(x, y, 50, paint);
        holder.unlockCanvasAndPost(canvas);
}
private class MyThread extends Thread {
        @Override
        public void run( ) {
                for (int i = 0; i < 40; i++) {
                        y += dy;
                        drawCanvas( );
                        try {
                                Thread.sleep(40);
                        } catch (InterruptedException e) {
                        }
                }
        }
}
@Override
public void surfaceChanged(SurfaceHolder holder, int format, int width,
```

```
        int height) {
    }

    @Override
    public void surfaceCreated(SurfaceHolder holder) {
        // drawCanvas( );
        new MyThread( ).start( );
    }

    @Override
    public void surfaceDestroyed(SurfaceHolder holder) {
    }
}
```

其效果是一个几何圆会从上至下落下，每次只变动更新Rect类别对象的区域而已。如下图：

## 7-4 以自定义View来实现手写签名App范例实作

直接就在这个单元来开发一个像样的App，也就是提供用户可以以手势来画出签名文件，并且可以进行存盘，主要是使用自定义View的方式来处理，外观就和7-1小节中一开始所规划的版面一样，先很快说明将前期工作准备到位，就可以开始进行开发。

## 7-4-1 前期准备

Project Name: BradSignature
Package Name: tw.brad.apps.BradSignature
Main Activity: MainActivity.java
Layout: activity_main.xml

整体操作风格是以横向握持为主,所以直接针对该Application设定为横向。

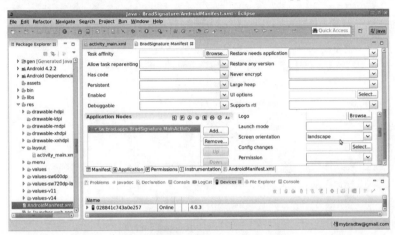

版面配置文件:activity_main.xml。

```
<LinearLayout xmlns:android="http://schemas.android.com/apk/res/android"
    android:layout_width="match_parent"
    android:layout_height="match_parent"
    android:orientation="horizontal" >

    <LinearLayout
        android:layout_width="match_parent"
        android:layout_height="match_parent"
        android:layout_weight="4"
        android:background="#1100ff00"
        android:orientation="vertical" >

        <Button
            android:id="@+id/clear"
            android:layout_width="match_parent"
            android:layout_height="wrap_content"
            android:layout_weight="1"
            android:text="Clear" />
```

```xml
<Button
    android:id="@+id/undo"
    android:layout_width="match_parent"
    android:layout_height="wrap_content"
    android:layout_weight="1"
    android:text="Undo" />

<Button
    android:id="@+id/redo"
    android:layout_width="match_parent"
    android:layout_height="wrap_content"
    android:layout_weight="1"
    android:text="Redo" />

<Button
    android:id="@+id/chcolor"
    android:layout_width="match_parent"
    android:layout_height="wrap_content"
    android:layout_weight="1"
    android:text="Color" />

<Button
    android:id="@+id/chsize"
    android:layout_width="match_parent"
    android:layout_height="wrap_content"
    android:layout_weight="1"
    android:text="Size" />
    </LinearLayout>
    <LinearLayout
        android:layout_width="match_parent"
        android:layout_height="match_parent"
        android:layout_weight="1"
        android:orientation="vertical" >
        <tw.brad.apps.BradSignature.PaintView
            android:id="@+id/pview"
            android:layout_width="match_parent"
            android:layout_height="match_parent"
            android:background="#44ffff00" />
    </LinearLayout>
</LinearLayout>
```

目前显示结果如7-1小节的呈现效果。

## ■ 7-4-2　开始处理签名的手势侦测处理

要画出线条的动作是以调用drawLine( )方法，因为需要一个Paint对象实体来处理画笔，先在建构式中事先设定好。

再来就是Override自定义View的onTouchEvent( )，并将其传回值设定为true，目的是希望当用户的一次触摸后滑动的所有点都可以触发取得。

目前的MainActivity.java如下：

```java
package tw.brad.apps.BradSignature;

import android.content.Context;
import android.graphics.Canvas;
import android.graphics.Color;
import android.graphics.Paint;
import android.util.AttributeSet;
import android.view.GestureDetector;
import android.view.MotionEvent;
import android.view.View;

public class PaintView extends View {
    private Paint paintLine;

    public PaintView(Context context, AttributeSet attrs) {
        super(context, attrs);

        // 设定画笔
        paintLine = new Paint( );
        paintLine.setColor(Color.GREEN);
        paintLine.setStrokeWidth(4);
    }

    @Override
    public boolean onTouchEvent(MotionEvent event) {
        return true;
    }

    @Override
    protected void onDraw(Canvas canvas) {
    }
}
```

接着准备一个数据结构来存放所有经过触摸点的线条对象，通常可以使用java.util.LinkedList类别对象实体来处理。而其元素泛型为java.util.HashMap的对象实体，以String为Key，其value为Float。

而用户的触摸动作：

• 开始触摸：首次执行onTouchEvent( )，要建构出线条的数据结构对象实体，放入第一触摸点。

• 开始滑动：放入第二及之后的触摸点，开始针对目前线条数据重新绘制。

• 离开屏幕：onTouchEvent( )方法侦测到离开屏幕，也就是MontionEvent的对象方法getAction( )的传回值，如果等于MotionEvent.ACTION_UP，就是离开屏幕的侦测。

修改如下：

```java
package tw.brad.apps.BradSignature;

import java.util.HashMap;
import java.util.LinkedList;

import android.content.Context;
import android.graphics.Canvas;
import android.graphics.Color;
import android.graphics.Paint;
import android.util.AttributeSet;
import android.view.MotionEvent;
import android.view.View;

public class PaintView extends View {
    private Paint paintLine;
    private LinkedList<HashMap<String,Float>> line;

    public PaintView(Context context, AttributeSet attrs) {
        super(context, attrs);

        // 设定画笔
        paintLine = new Paint( );
        paintLine.setColor(Color.GREEN);
        paintLine.setStrokeWidth(4);

        // 初始建构线条数据结构对象
        line = new LinkedList<HashMap<String,Float>>( );
    }
```

```java
@Override
public boolean onTouchEvent(MotionEvent event) {
    if (event.getAction( ) == MotionEvent.ACTION_UP){
        // 离开屏幕
        line.clear( );
    }else {
        // 触摸开始
        HashMap<String,Float> point = new HashMap<String, Float>( );
        point.put("x", event.getX( ));
        point.put("y", event.getY( ));
        line.add(point);
        postInvalidate( );
    }
    return true;
}

@Override
protected void onDraw(Canvas canvas) {
    if (line.size( )>1){
        // 至少有两点才需要画出线段
        for (int i=1; i<line.size( ); i++){
            // i = 1表示从第二点开始处理
            HashMap<String,Float> point1 = line.get(i-1);
            // 前一点
            HashMap<String,Float> point2 = line.get(i);
            // 目前点
            canvas.drawLine(point1.get("x"), point1.get("y"),
                point2.get("x"), point2.get("y"),
                paintLine);
        }
    }
}
```

测试执行之后发现，可以画出一条线，但是当用户离开触摸屏幕后，该线条就消失了，但是又可以继续画出下一条线。原因就是从头到尾只有使用一个LinkedList来放置一条线的数据，而触摸离开屏幕时为了要画下一条线，而将之前的LinkedList的数据清除，才会有上述状况发生。

再来就准备一个多条线的数据结构,也是LinkedList,但是其存放元素为刚刚的线条的LinkedList,因此会是如下的声明处理:

```
private LinkedList<LinkedList<HashMap<String,Float>>> lines;
```

而在建构式中:

```
lines = new LinkedList<LinkedList<HashMap<String,Float>>>( );
```

再来就是调整逻辑结构。

设定一个boolean变量来判断是否处在触摸屏幕状态,一开始其值为false,当触摸开始,将其值设定为true,也同时建构一条线的数据结构对象,并加进lines的数据结构之中;而开始滑动的判断就是isTouching值为true的状态,则是从lines中取出最后的线条继续增加触摸点数据。

```java
package tw.brad.apps.BradSignature;

import java.util.HashMap;
import java.util.LinkedList;

import android.content.Context;
import android.graphics.Canvas;
import android.graphics.Color;
import android.graphics.Paint;
import android.util.AttributeSet;
import android.view.MotionEvent;
import android.view.View;

public class PaintView extends View {
    private Paint paintLine;
    private LinkedList<HashMap<String,Float>> line;
    private LinkedList<LinkedList<HashMap<String,Float>>> lines;
    private boolean isTouching;

    public PaintView(Context context, AttributeSet attrs) {
        super(context, attrs);

        // 设定画笔
        paintLine = new Paint( );
        paintLine.setColor(Color.GREEN);
        paintLine.setStrokeWidth(4);
```

```java
        // 初始建构线条数据结构对象
        lines = new LinkedList<LinkedList<HashMap<String,Float>>>();
        isTouching = false;
}

@Override
public boolean onTouchEvent(MotionEvent event) {
    if (event.getAction( ) == MotionEvent.ACTION_UP){
        // 离开屏幕
        isTouching = false;
    }else {
        LinkedList<HashMap<String,Float>> line;
            if (!isTouching){
                    // 触摸开始
                    line = new LinkedList<HashMap<String,Float>>();
                    lines.add(line);
                    isTouching = true;
            }else{
                    // 开始滑动
                    line = lines.getLast( );
            }
            HashMap<String,Float> point = new HashMap<String, Float>();
            point.put("x", event.getX( ));
            point.put("y", event.getY( ));
            line.add(point);
            postInvalidate( );
    }
    return true;
}
@Override
protected void onDraw(Canvas canvas) {
        for (LinkedList<HashMap<String,Float>> line: lines){
                if (line.size( )>1){
                        // 至少有两点才需要画出线段
                        for (int i=1; i<line.size( ); i++){
                                // i = 1表示从第二点开始处理
```

```
                HashMap<String,Float> point1 = line.get(i-1);
                    // 前一点
                HashMap<String,Float> point2 = line.get(i);
                    // 目前点
                canvas.drawLine(point1.get("x"), point1.get("y"),
                point2.get("x"), point2.get("y"),
                paintLine);
                }
            }
        }
    }
}
```

终于可以画出多条线段，如下图：

### 7-4-3　处理外部功能

进行到要处理左侧功能键的功能，先从清除功能下手。清除的功能对应到程序中，就是将 lines 中所有线段的数据清除，并要求重新绘制即可。此时开发一个自定义 View 的方法来负责这件事。

```
void clearDraw( ){
    lines.clear( );
    postInvalidate( );
}
```

回到 MainActivity.java 中设定 Button 调用自定义 View 的 clearDraw( ) 方法。

MainActivity.java

```
package tw.brad.apps.BradSignature;
import android.app.Activity;
```

```java
import android.os.Bundle;
import android.view.View;
import android.view.View.OnClickListener;
public class MainActivity extends Activity {
    private PaintView pview;
    private View clear;
    @Override
    protected void onCreate(Bundle savedInstanceState) {
        super.onCreate(savedInstanceState);
        setContentView(R.layout.activity_main);
        pview = (PaintView)findViewById(R.id.pview);
        clear = findViewById(R.id.clear);
        clear.setOnClickListener(new OnClickListener( ) {
            @Override
            public void onClick(View v) {
                pview.clearDraw( );
            }
        });
    }
}
```

再来处理Undo/Redo功能，Undo相对PaintView程序中增加一个对象方法，将lines中的最后一条line元素进行移除，然后记得重新绘制即可；而redo是要将刚刚移除的line再度叫回来。所以配合这两个功能，增设一个与lines一样的对象变量recyle，视为移除line对象的资源回收桶。从lines移除的line放在recyle；而需要Redo功能时，就从recyle中的line对象移除，放回lines中即可。

PaintView.java中先声明定义：

```java
private LinkedList<LinkedList<HashMap<String, Float>>> recyle;
```

建构时机与lines相同即可：

```java
recyle = new LinkedList<LinkedList<HashMap<String, Float>>>( );
```

开发Undo/Redo的对象方法：

```java
// Undo功能
```

```
void undoDraw( ){
    if (lines.size( )>0){
        recycle.add(lines.removeLast( ));
        postInvalidate( );
    }
}

// Redo功能
void redoDraw( ){
    if (recycle.size( )>0){
        lines.add(recycle.removeLast( ));
        postInvalidate( );
    }
}
```

回到MainActivity.java中使用功能：

```
undo = findViewById(R.id.undo);
undo.setOnClickListener(new OnClickListener( ) {
    @Override
    public void onClick(View v) {
        pview.undoDraw( );
    }
});
redo = findViewById(R.id.redo);
redo.setOnClickListener(new OnClickListener( ) {
    @Override
    public void onClick(View v) {
        pview.redoDraw( );
    }
});
```

到此的练习都是相当简单的，看到如何处理自定义View的模式，后面其他功能在后续的章节中继续完成。

# 08 Chapter

## 第8课　数据存取

8-1　偏好设定

8-2　内部文件存取机制

8-3　外部文件存取

8-4　移动装置数据库处理机制SQLite

8-5　应用App资源中的数据存取数据：以游戏关卡数据处理为例

## 8-1 偏好设定

在应用程序中，通常会将用户个人化的相关设定进行储存，以利于下次执行时直接调用使用，用户不会感觉到每次执行都和首次执行一样，还要进行相关设定。例如用户名称、音乐音效开启状态或是记录游戏关卡状态等。

android.content.SharedPreferences 类别是用来针对基本类型的 key-value 成对的名称数据进行存取机制的架构，利用该类别所建构的对象可以存放的基本类型有 boolean、float、int、long 及字符串对象类型的数据。

### ■ 8-1-1 处理方式

通常在应用程序中取得 android.content.SharedPreferences 类别对象的方式有两种。

① 调用 getSharedPreferences(String name，int mode)。通常适用于应用程序中会使用多个数据储存文件。调用该方法必须传入两个参数：

- 第一个参数项为自定义的储存数据文件名称，如果指定的文件名不存在，则会在实际进行存取的动作同时自动建立。
- 第二个参数为指定的操作模式，通常默认为 MODE_PRIVATE(0)。

② 调用 getPreferences(int mode)。通常适用于应用程序中会使用多个数据储存文件。所以与上个方法不同的地方就是不需要指定自定义的储存数据文件名称，只需传入一个参数项，参数为指定的操作模式，通常默认为 MODE_PRIVATE(0)。

两种方法都可以在继承 Activity 类别的子类别中直接调用，就可以取得 SharedPreferences 的类别对象。

### ■ 8-1-2 基本处理程序

储存数据的基本程序如下：

① 调用 SharedPreferences 的类别对象的 edit( ) 方法，取得传回的 Shared Preferences.Editor 对象。

② 调用 sharedPreferences.Editor 对象的 putXxx(String key,Xxx value) 方法，将数据进行储存。Xxx 表示基本类型的 Boolean、Float、Int、Long 及 String。

③ 最后调用 SharedPreferences.Editor 对象的 commit( ) 方法，将数据存放。

④ 如果调用 SharedPreferences.Editor 对象的 clear( ) 方法，将会清除数据。

### ■ 8-1-3 范例说明

先声明 SharedPreferences 和 SharedPreferences.Editor 两个对象变量。

```
private SharedPreferences sp;
private SharedPreferences.Editor editor;
```

在程序中建构:

```
sp = getSharedPreferences("games", MODE_PRIVATE);
editor = sp.edit( );
```

指定读取文件名称为games,读取偏好设定数据。

```
private void readPreferences( ){
    String user = sp.getString("user", "nobody");
    boolean sound = sp.getBoolean("sound", false);
    int stage = sp.getInt("stage", 0);
    info.setText("User:" + user + "\n" +
                 "Sound:" + (sound?"On":"Off") + "\n" +
                 "Stage:" + stage);
}
```

上述程序代码中:

• 指定读取Key值为"user"的字符串数据,调用getString( ),第二个参数项为当读不到指定Key值时使用的默认值"nobody"。

• 指定读取Key值为"sound"的boolean数据,调用getBoolean( ),第二个参数项为当读不到指定Key值时使用的默认值false。

• 指定读取Key值为"stage"的字符串数据,调用getInt( ),第二个参数项为当读不到指定Key值时使用的默认值0。

储存偏好设定数据:

```
private void savePreferences( ){
    // 清除数据
    //editor.clear( );
    editor.putBoolean("sound", true);
    editor.putInt("stage", 4);
    editor.putString("user", "brad");
    editor.commit( );
    Toast.makeText(this, "Save OK", Toast.LENGTH_SHORT).show( );
}
```

上述程序代码中，由editor对象实体进行数据储存：
- 调用putBoolean( )，指定Key字符串值存放boolean数据；
- 调用putInt( )，指定Key字符串值存放int数据；
- 调用putString( )，指定Key字符串值存放String数据；
- 最后调用commit( )方法进行实体文件写出。

当尚未进行储存设定数据，而先执行读取偏好设定数据时，将会以默认数据进行给值，也就是所谓的应用程序初始默认值。呈现画面如下图所示。

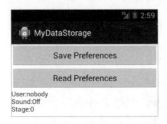

而当进行数据设定储存之后，将可以在File Explorer的窗口中观察，数据文件被放在/data/data/<Package-Name>/shared_prefs/目录下的games.xml，也就是说，程序中只需要指定数据文件名，将会被自动附加扩展名为.xml，其内容将会如下：

```
<?xml version='1.0' encoding='utf-8' standalone='yes' ?>
<map>
<int name="stage" value="4" />
<string name="user">brad</string>
<boolean name="sound" value="true" />
</map>
```

再度读取数据，将会正确读到如下图所示。

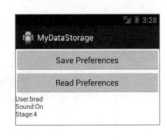

> 这样的数据内容，当用户在应用程序管理中，进行数据清除（Clear Data）后，就会将该文件删除，或是解除安装（Uninstall）也是。

## 8-1-4 完整范例

① activity_main.xml

```xml
<LinearLayout xmlns:android="http://schemas.android.com/apk/res/android"
    android:layout_width="match_parent"
    android:layout_height="match_parent"
    android:orientation="vertical"
    >

<Button
    android:id="@+id/save1"
    android:layout_width="match_parent"
    android:layout_height="wrap_content"
    android:text="Save Preferences"
    />
<Button
    android:id="@+id/read1"
    android:layout_width="match_parent"
    android:layout_height="wrap_content"
    android:text="Read Preferences"
    />
<TextView
    android:id="@+id/info"
    android:layout_width="wrap_content"
    android:layout_height="wrap_content"
    />
</LinearLayout>
```

② MainActivity.java

```java
package tw.brad.android.book.MyDataStorage;

import android.app.Activity;
import android.content.SharedPreferences;
import android.os.Bundle;
import android.view.View;
import android.view.View.OnClickListener;
import android.widget.TextView;
```

```java
import android.widget.Toast;
public class MainActivity extends Activity {
    private View save1, read1;
    private TextView info;
    private SharedPreferences sp;
    private SharedPreferences.Editor editor;

    @Override
    protected void onCreate(Bundle savedInstanceState) {
        super.onCreate(savedInstanceState);
        setContentView(R.layout.activity_main);

        sp = getSharedPreferences("games", MODE_PRIVATE);
        editor = sp.edit( );

        info = (TextView) findViewById(R.id.info);
        save1 = findViewById(R.id.save1);
        save1.setOnClickListener(new OnClickListener( ) {
            @Override
            public void onClick(View v) {
                savePreferences( );
            }
        });
        read1 = findViewById(R.id.read1);
        read1.setOnClickListener(new OnClickListener( ) {
            @Override
            public void onClick(View v) {
                readPreferences( );
            }
        });
    }

    private void savePreferences( ) {
        // 清除数据
        // editor.clear( );
        editor.putBoolean("sound", true);
        editor.putInt("stage", 4);
        editor.putString("user", "brad");
        editor.commit( );
```

```
            Toast.makeText(this, "Save OK", Toast.LENGTH_SHORT).show( );
    }
    private void readPreferences( ) {
        String user = sp.getString("user", "nobody");
        boolean sound = sp.getBoolean("sound", false);
        int stage = sp.getInt("stage", 0);
        info.setText("User:"+ user +"\n"+"Sound:"+ (sound ? "On":"Off")
            + "\n" + "Stage:" + stage);
    }
}
```

## 8-2 内部文件存取机制

### ■ 8-2-1 使用观念

在应用程序执行中,需要针对用户相关的文件进行存取,而这种类型的文件的特性,通常是仅针对应用程序使用而已,例如程序中的对象序列化文件,当下次执行该应用程序的时候,或是用户手动暂停后继续执行,可以读取该对象解串行化文件继续之前的执行过程等。主要的观念在于该类型的文件是配合应用程序才有作用,当用户卸载之后,这类型的文件就没有存在的意义,那就非常适用使用内部文件访问机制。千万别与应用程序的照相功能或是录音功能所产生的文件访问机制搞混,因为相片文件或是其他类似形式文件,用户利用应用程序产生之后,还是可以通过其他应用程序进行存取,那种文件访问机制就是属于外部文件存取,而文件存放在SDCard。本小节的文件与上一节的偏好设定一样,是存放在内存空间。

### ■ 8-2-2 写出基本程序

建立及写出内部文件到装置的储存空间的基本程序:
• 调用Activity的openFileOutput( )方法,传回一个java.io.FileOutputStream对象实体。
• 以java.io.FileOutputStream对象实体的write( )方法进行数据写入装置的储存空间。
• 写入完成之后,调用java.io.FileOutputStream对象实体的close( )方法以关闭文件输出串流。
建立一个文件输出串流:

```java
private void saveInnerFile( ){
    FileOutputStream fout;
    try {
        fout = openFileOutput("MyData.txt", MODE_PRIVATE);
        fout.flush( ); // 清除内存缓冲
        fout.close( ); // 关闭串流
    } catch (FileNotFoundException e)
    {     e.printStackTrace( );
    } catch (IOException e) {
        e.printStackTrace( );
    }
}
```

调用openFileOutput( )方法，传递两个参数：
① 文件名。
② 开启模式
　　• MODE_PRIVATE：每次写出会删除原来内容。
　　• MODE_APPEND：每次写出会保留原内容，而从文件尾端开始写出。
　　以上方法虽然没有写入任何数据，一旦执行之后，就已经会产生指定的文件/data/data/<Package-Name>/files/MyData.txt，文件大小为0。
　　接着进行文字数据写出，调用FileOutputStream对象实体的write( )方法，将欲写出的数据转成byte数组形式，传递参数写出即可。

```java
private void saveInnerFile( ) {
    FileOutputStream fout;
    String data = "Hello, World!\n 我是赵令文\n";
    try {
        fout = openFileOutput("MyData.txt", MODE_PRIVATE);
        fout.write(data.getBytes( ));
        fout.flush( );
        fout.close( );
    } catch (FileNotFoundException e) {
        e.printStackTrace( );
    } catch (IOException e) {
        e.printStackTrace( );
    }
}
```

可以从 File Explorer 观察到文件的变化，如下图所示。

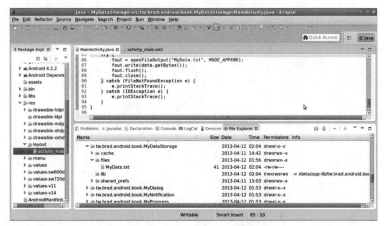

## 8-2-3 读入基本程序

从装置储存空间读取内部文件的基本程序：

① 调用 Activity 的 openFileInput( ) 方法，传回一个 java.io.FileInputStream 对象实体。

② 以 java.io.FileInputStream 对象实体的 read( ) 方法从装置的储存空间读入一个字节的数据。

③ 写入完成之后，调用 java.io.FileInputStream 对象实体的 close( ) 方法以关闭文件输入串流。

建立一个输入串流：

```java
private void readInnerFile( ) {
    FileInputStream fin;

    try {
        fin = openFileInput("MyData.txt");

        fin.close( );
    } catch (FileNotFoundException e) {
        e.printStackTrace( );
    } catch (IOException e) {
        e.printStackTrace( );
    }
}
```

调用 openFileInput( )，传入指定文件名字符串即可。文件名直接就是相对 /data/data/<Package-Name>/files 文件夹进行指定。

以下范例将文件内容显示在 TextView 中：

```java
private void readInnerFile( ) {
    FileInputStream fin;
    try {
        fin = openFileInput("MyData.txt");
        BufferedReader reader =
                new BufferedReader(new InputStreamReader(fin));
        String line;
        info.setText("");
        while ((line = reader.readLine( )) != null){
            info.append(line + "\n");
        }
        reader.close( );
        fin.close( );
    } catch (FileNotFoundException e) {
        e.printStackTrace( );
    } catch (IOException e) {
        e.printStackTrace( );
    }
}
```

因为文件内容为一般文本文件，所以先以 InputStreamReader 对象将其转换为 Reader 串流对象，再将其串接在 BufferedReader 串流对象，以方便调用 readLine( ) 方法，一次读入一列文字数据。

执行后显示于移动装置如下图：

这样的数据内容，当用户在应用程序管理中，进行数据清除（Clear Data）后，就会将该文件删除，或是卸载（Uninstall）也是。

## 8-3 外部文件存取

外部文件存取就是针对SDCard的文件系统进行存取。

### 8-3-1　SDCard文件系统基本概念

- 与Linux文件系统相同的单根文件系统。
- 文件目录名称大小写严格区分，B与b是不同的。
- SDCard通常是挂载在/mnt/sdcard目录下，最好以API来进行判断。
- 数据文件写出必须开启特定写出权限。

判断用户的移动装置是否有SDCard：调用android.os.Environment的static方法isExternalStorageRemovable( )即可判断是否有挂载SDCard。传回true表示没有挂载；传回false则表示已经挂载。

```
if (!Environment.isExternalStorageRemovable( )){
    info.setText("SDCard Mounted");
}else {
    info.setText("SDCard Removed");
}
```

上述方式至少要在API Level 9+，若是API Level 8以前的移动装置，则可以调用Environment.getExternalStorageState( )传回字符串判断：
- "removed"：已经移除SDCard。
- "mounted"：已经挂载SDCard。

如果以仿真器进行测试，记得将仿真器的SDCard设定适当的大小，通常不需要设定太大，笔者大约设定成128MB就足够了。

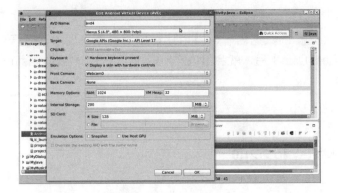

## 8-3-2 判断SDCard的挂载点（Mount Point）

调用Environment.getExternalStorageDirectory( )传回SDCard所在的目录的File对象实体，可以在调用File对象实体的getAbsolutePath( )方法传回字串内容。

```
info.setText("SDCard Mounted:" +
    Environment.getExternalStorageDirectory( ).getAbsolutePath( ));
```

其他特定目录都是针对SDCard所在目录的子目录：
- Environment.DIRECTORY_ALARMS。
- Environment.DIRECTORY_DCIM。
- Environment.DIRECTORY_DOWNLOADS。
- Environment.DIRECTORY_MOVIES。
- Environment.DIRECTORY_MUSIC。
- Environment.DIRECTORY_NOTIFICATIONS。
- Environment.DIRECTORY_PICTURES。
- Environment.DIRECTORY_PODCASTS。
- Environment.DIRECTORY_RINGTONES。

## 8-3-3 应用程序文件应该在哪里

应用程序文件存取原则应该是放在适当的专属子目录底下，如果开发者认定文件内容特性是跟着应用程序一起"存亡"的话，则应该放在:/mnt/sdcard/Android/data/<Package-Name>/的子目录下。如此一来，当用户卸载应用程序之后，该子目录下的所有内容也将一并删除，不留痕迹。而放在其他子目录下的文件则不受卸载应用程序的影响。

## 8-3-4 开启写出数据的权限

如果没有开启写出权限，即使程序没有任何逻辑错误，也将会在用户的运行时间内不因异常而中止执行。以下就进行用户权限开启：

开启项目目录下的AndroidManifest.xml文件，点选Permissions的页面。

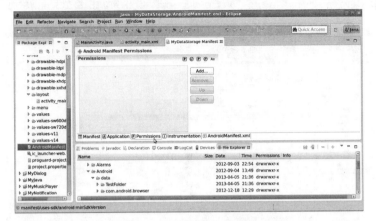

按下"Add"之后，如下图点选"User Permission"。

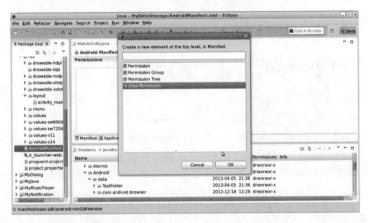

在下图中右侧以下拉式菜单点选出"android.permission.WRITE_EXTERNAL_STORAGE"并按下"Ctrl+S"保存即可。

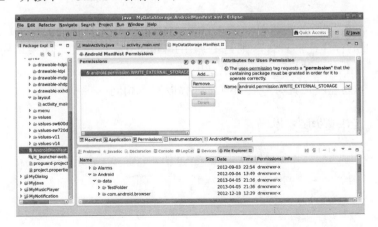

或是直接在 AndroidManifest.xml 页签中直接输入：

```
<uses-permission android:name="android.permission.WRITE_EXTERNAL_STORAGE"/>
```

### 8-3-5 开始进行程序开发

声明 SDCard 与应用程序的子目录 File 对象变量：

```
private File sdroot, approot;
```

取得 SDCard 的 File 对象实体，再由其建构应用程序专属子目录的 File 对象实体。

```
sdroot = Environment.getExternalStorageDirectory( );
approot = new File(sdroot, "Android/data/" + getPackageName( ));
if (!approot.exists( )){
    approot.mkdirs( );
}
```

因为一开始用户首次执行时，可能并不存在应用程序专属子目录，因此先以 exists( ) 判断是否存在，若否，则马上调用 mkdirs( ) 进行建立。如果调用 mkdir( )，则可能会因为不存在父目录而无法建立出来。

接下来的处理模式就与一般 Java 文件存取完全相同。

### 8-3-6 写出数据文件

```
private void saveSDCardFile( ){
    String data = "Hello, SDCard";
    FileOutputStream fout;
```

```
    try {
         fout = new FileOutputStream(new File(approot, "brad.data"));
         fout.write(data.getBytes( ));
         fout.flush( );
         fout.close( );
    } catch (FileNotFoundException e) {
         e.printStackTrace( );
    } catch (IOException e) {
         e.printStackTrace( );
    }
}
```

则将会写出数据于 /mnt/sdcard/Android/data/<Package-Name>/brad.data 如下图所示。

### 8-3-7 读入数据文件

```
private void readSDCardFile( ){
    FileInputStream fin;
    try {
         info.setText("");
         fin = new FileInputStream(new File(approot, "brad.data"));
         BufferedReader reader =
              new BufferedReader(
                   new InputStreamReader(fin));
```

```
            String line;
            info.setText("");
            while ((line = reader.readLine( )) != null) {
                info.append(line + "\n");
            }
            reader.close( );
            fin.close( );
    } catch (FileNotFoundException e) {
            e.printStackTrace( );
    } catch (IOException e) {
            e.printStackTrace( );
    }
}
```

## 8-4 移动装置数据库处理机制SQLite

开发应用程序运用的数据库系统来处理大量的数据，使用 SQL 查询语法可以迅速地过滤出想要的数据，这应该是数据库系统有别于一般文件输入输出的优势。一样是存放数据，但是放在数据库中的数据比较具有使用上的意义，而以一般文件存放数据的缺点就是查询过滤的复杂性与效能性远低于数据库。Android 完全支持 SQLite 数据库系统。

### 8-4-1 建立数据库的辅助类别对象

自行开发一个子类别继承 SQLiteOpenHelper 类别，用来建立数据库及数据表。实作方式如下：

- Override: onCreate(SQLiteDatabase db)。
- Override: onUpgrade(SQLiteDatabase db，int oldVersion，int newVersion)。
- 定义建构式：MyDBHelper(Context context，String dbname，CursorFactory factory，int version)。

### 8-4-2 预先处理模式

以实际范例来说明处理模式，假设 App 中会使用到数据库来存放客户相关数据，因此会有一个数据表存在于数据库中，该数据表的结构大致如下：

- _id：整数形式；主键；自动递增。
- name：文字形式。
- tel：文字形式。
- birthday：日期形式。

依据SQLite的语法来建立该数据表（cust）：

```
CREATE TABLE cust (_id INTEGER PRIMARY KEY AUTOINCREMENT, name TEXT, tel TEXT, birthday DATE)
```

因此开发出一个自定义 SQLiteOpenHelper类别如下（MyDBHelper）：

```java
package tw.brad.android.book.MyDataStorage;

import android.content.Context;
import android.database.sqlite.SQLiteDatabase;
import android.database.sqlite.SQLiteDatabase.CursorFactory;
import android.database.sqlite.SQLiteOpenHelper;

public class MyDBHelper extends SQLiteOpenHelper {
    private static final String createCustTable =
            "CREATE TABLE cust (_id INTEGER PRIMARY KEY AUTO_INCREMENT," +
            "name TEXT, tel TEXT, birthday DATE)";

    public MyDBHelper(Context context, String dbname, CursorFactory factory,
            int version) {
        super(context, dbname, factory, version);
    }

    @Override
    public void onCreate(SQLiteDatabase db) {
        db.execSQL(createCustTable);
    }

    @Override
    public void onUpgrade(SQLiteDatabase db, int oldVersion, int newVersion) {
    }
}
```

此时就可以开发到主程序进行开发。先进行声明：

```java
private MyDBHelper dbhelper;
private SQLiteDatabase db;
```

接着调用 dbhelper 对象方法 getReadableDatabase( ) 或是 getWritableDatabase( ) 回传 SQLiteDatabase 对象存放在 db 变量。两者的差异在于用户的储存空间不足的情况之下，getReadableDatabase( ) 仍然可以开启使用，只是处在只读的模式，一旦用户释放出足够储存空间，则又可以开始进行读写模式；而 getWritableDatabase( ) 则在用户的储存空间不足的情况之下，直接抛出 Exception，开发者将会针对该 Exception 进行 try...catch 的开发结构来处理。

以下建立出数据库名称为"brad"：

```
dbhelper = new MyDBHelper(this, "brad", null, 1);
db = dbhelper.getReadableDatabase( );
//db = dbhelper.getWritableDatabase( );
```

而数据库的关闭时机，以下放在该 Activity 结束时：

```
@Override
public void finish( ) {
    if (db.isOpen( )){
        db.close( );
    }
    super.finish( );
}
```

### 8-4-3　简单查询数据

先来开发撰写最简单也最常用的查询语法：SELETE * FROM cust

```
private void querySQLiteData( ) {
    Cursor c = db.query("cust", null, null, null, null, null, null);
    info.setText("笔数:" + c.getCount( ) + "\n");
    while (c.moveToNext( )){
    info.append(c.getString(c.getColumnIndex("_id")) + ":" +
            c.getString(c.getColumnIndex("name")) + ":" +
            c.getString(c.getColumnIndex("tel")) + ":" +
            c.getString(c.getColumnIndex("birthday")) + "\n");
    }
}
```

- 调用 SQLiteDatabase 对象 db 的 query( )，传递第一个参数为数据表名称，其余先暂时设定为 null，共有 6 个 null。

- query( ) 方法传回 Cursor 对象实体。
- 调用 Cursor 对象实体的 moveToNext( )，使查询指针往下一笔数据移动，当没有任何数据时，将会传回 false(boolean)。
- 通过 Cursor 对象实体的 getCount( ) 方法传回查询结果的数据笔数。
- 通过 Cursor 对象实体的 getColumnIndex( ) 指定查询域名，传回其传回结果字段的 index。
- 再调用 Cursor 对象实体的 getString( )，指定字段 index，传回该笔数据内容。

### ■ 8-4-4 新增数据

直接调用 SQLiteDatabase 对象实体的 insert( ) 方法，传递三个参数：
- 数据表名称。
- null：通常设定为 null。
- 数据内容。

```
private void insertSQLiteData( ){
    ContentValues values = new ContentValues( );
    values.put("name", "test" + (int)(Math.random( )*100));
    values.put("tel", "0999-123456");
    values.put("birthday", "1999-12-12");
    db.insert("cust", null, values);
}
```

数据内容以 ContentValues 对象来实作，调用其 put(Key, Value)，将字段字符串当作其 Key，而其数据值放在第二的参数传递。上例中建立了 name 字段为 "test"+0 ~ 99 的数字数据。

### ■ 8-4-5 删除数据

直接调用 SQLiteDatabase 对象实体的 delete( ) 方法，传递三个参数：
- 数据表名称。
- 查询条件式。
- 查询条件值字符串数组。

以下列的 SQL 语法为例：

```
DELETE FROM cust WHERE name like 'test%'
```

则写成如下：

```
private void deleteSQLiteData( ) {
    // db.delete("cust", "name like ?", new String[]{"test_"});
    db.delete("cust", "name like ?", new String[] { "test%" });
}
```

批注列用来删除 test0 ～ test9 而已。查询条件式中的问号可以多个，而多个问号依照由左至右的顺序，一对一地对应到查询条件值字符串数组的值。

### ■ 8-4-6 修改数据

直接调用 SQLiteDatabase 对象实体的 update( ) 方法，传递四个参数：
- 数据表名称。
- 修改的数据内容。
- 查询条件式。
- 查询条件值字符串数组。

以下列的 SQL 语法为例：

```
UPDATE cust SET name='赵令文',birthday='1999-01-01' WHERE name='Brad'
```

实作如下：

```
private void updateSQLiteData( ) {
    ContentValues values = new ContentValues( );
    values.put("name", "赵令文");
    values.put("birthday", "1999-01-01");
    db.update("cust", values, "name = ?", new String[] { "Brad" });
}
```

### ■ 8-4-7 进一步了解查询方式

再来回头看看查询的参数项：
① 数据表名称。
② 查询字段字符串数组。
③ 查询条件式。
④ 查询条件值字符串数组。
⑤ GROUP BY 字符串语法。
⑥ Having 字符串语法。
⑦ ORDER BY 字符串语法。

以下列SQL为例：

```
SELETE _id, name, birthday WHERE _id>4 ORDER BY _id
```

实作如下：

```
Cursor c = db.query("cust",
                    new String[]{"_id", "name", "birthday"}
                    , "_id > ?",
                    new String[]{"4"},
                    null,
                    null,
                    "_id");
```

## 8-5 应用App资源中的数据存取数据：以游戏关卡数据处理为例

应用程序中事先设定好的数据内容，提供应用程序在执行中存取使用，给用户在不同的情境执行，可以使用本小节的处理模式。例如游戏App中的关卡地图数据，或是应用程序中给用户不同的操作环境相关数据。

数据存放在项目架构下的res/目录，自行建立出res/raw/的子目录，将数据相关文件放置于该子目录下，如下图所示：

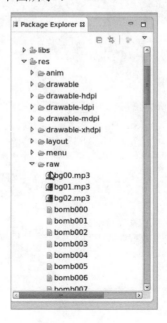

提醒：子目录名称为全小写raw，不得任意修改，修改原则是依照资源名称规则。

而在程序中将会事先定义一个int[]数组存放所需要用到的部分：关卡数据。

```
private int[] leveldata = {
    R.raw.bomb000, R.raw.bomb001, R.raw.bomb002, R.raw.bomb003,
    R.raw.bomb004, R.raw.bomb005, R.raw.bomb006, R.raw.bomb007};
```

### ■ 8-5-1　定义数据

而数据内容为自行定义，本例中定义如下：
- 一列就是一个关卡地图中的显示内容一列。
- 0：不显示。
- 1：显示砖块。
- 2：显示岩石。
- 7：显示敌人。
- 8：显示主角。
- 9：显示目标物。

预计显示如下游戏画面：

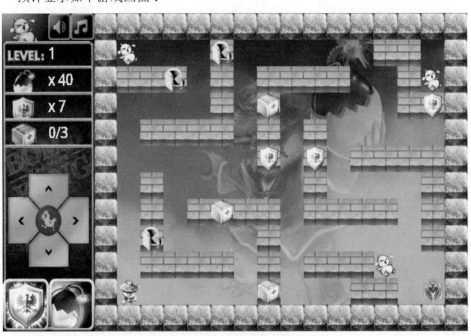

右大半边就是整个关卡的地图,目前主角在左下方,敌人在右下方,有三个目标物要去取得,而散布在砖块中的宝物是在程序中以随机数生成,所以不在关卡中设定,以免玩家玩出密技,增加游戏变化的趣味性。

### ■ 8-5-2 读取数据文件

产生一般简单的文字文件内容如下:
bomb001

```
0,0,0,0,2,2,2,2,2,2,2,2,2,2,2,2,2,2,2,2
0,0,0,0,2,9,0,0,0,1,0,0,0,0,0,0,1,1,0,2
0,0,0,0,2,1,1,1,0,1,0,1,1,1,1,0,0,1,9,2
0,0,0,0,2,0,0,0,0,1,0,1,0,0,0,0,0,1,1,2
0,0,0,0,2,0,1,1,1,1,0,1,0,1,0,0,0,0,0,2
0,0,0,0,2,0,0,0,0,0,1,0,1,0,1,1,1,1,1,2
0,0,0,0,2,0,1,0,0,0,0,0,0,1,0,0,0,0,1,2
0,0,0,0,2,0,1,0,1,1,1,1,0,1,1,1,1,0,1,2
0,0,0,0,2,0,1,0,0,0,0,0,0,0,0,0,0,0,1,2
0,0,0,0,2,0,1,1,1,1,0,1,0,1,1,1,9,0,0,2
0,0,0,0,2,8,0,0,0,0,0,1,0,0,0,1,1,0,7,2
0,0,0,0,2,2,2,2,2,2,2,2,2,2,2,2,2,2,2,2
```

### ■ 8-5-3 程序中读取方式

通过调用 Context 的 getResources( ) 方法传回 Resources 对象实体,也就是用来存取资源数据的对象。再调用其 openRawResource( ),传递资源变量整数值,则将传回一个 InputStream 对象实体。

```
BufferedReader reader =new BufferedReader(
    new InputStreamReader(res.openRawResource(leveldata[n])));
```

之后进行数据解析,写入事先定义的二维数组 gamemap[][] 中。

```
String temp1;
String[] temp3;
int i = 0, j = 0, v;
while ((temp1 = reader.readLine( )) != null) {
    temp3 = temp1.split(",");
```

```
        for (String temp4 : temp3) {
            v = Integer.parse Int(temp4);
            gamemap[i][j] = v;
        }
    }
}
```

当然，以上仅用来说明读取项目资源的文件，解析数据方式非常多，读者可以考虑以XML方式作为数据格式。

# 第9课　因特网相关

9-1　网络接口及 IP Address

9-2　UDP 通信协议的数据存取

9-3　TCP 通信协议的数据存取

9-4　Http 通信协议的数据存取

9-5　WebView 使用

## 9-1 网络接口及IP Address

移动装置中网络装置的相关状态取得,是有助于存取网络相关服务的重要信息。用户的移动装置的网络联机状态,使用的网络服务形式,以及目前所使用的IP Address等,都会在往后的网络服务中使用。

因此,本小节先来进行移动装置上的网络相关信息的取得。

### ■ 9-1-1 装置的网络状态

android.net.NetworkInfo类别对象可以用来取得目前网络的联机状态,以及提供服务的形式。而该对象的取得是通过android.net.ConnectivityManager对象实体调用其getActiveNetworkInfo( )方法传回的。ConnectivityManager是由getSystemService( )方法传入Context.CONNECTIVITY_SERVICE得到的。

```
ConnectivityManager connMgr =
    (ConnectivityManager)getSystemService(CONNECTIVITY_SERVICE);
NetworkInfo networkInfo = connMgr.getActiveNetworkInfo( );
```

接着就可以由networkInfo对象实体调用isConnected( )方法传回boolean值来判断目前是否已经联机;调用其getTypeName( )方法传回联机服务为WIFI或是MOBILE。

```
if (networkInfo != null && networkInfo.isConnected( )) {
    sb.append("Type: " + networkInfo.getTypeName( ) + "\n");
} else {
    sb.append("Not Connect");
}
```

以上部分的使用调用,必须要开启android.permission.ACCESS_NETWORK_STATE的权限。在androidManifest.xml中加上:

```
<uses-permission android:name="android.permission.ACCESS_NETWORK_STATE"/>
```

### ■ 9-1-2 网络接口的IP Address

如果要继续以下的网络适配器信息的取得,则必须再开启android.permission.

INTERNET 的权限。

```
<uses-permission android:name="android.permission.INTERNET"/>
```

调用java.net.NetworkInterface 的 getNetworkInterfaces( )方法将传回 Enumeration <NetworkInterface>的对象实体，代表移动装置上目前所侦测到的网络接口。而每个网络接口上都可被绑定一组以上的 IP Address，所以可以用以下方式进行查找：

```
try {
    Enumeration<NetworkInterface> ifcs = NetworkInterface
            .getNetworkInterfaces( );
    while (ifcs.hasMoreElements( )) {
        NetworkInterface ifc = ifcs.nextElement( );
        sb.append("Interface: " + ifc.getDisplayName( ) + "\n");
        List<InterfaceAddress> ips = ifc.getInterfaceAddresses( );
        for (InterfaceAddress ip : ips) {
            sb.append("\tIP: " + ip.getAddress( ).getHostAddress( )
                    + "\n");
        }
    }
} catch (SocketException e) {
    sb.append("XXX");
}
```

实际测试后的结果如下：
① Android 2.3 手机装置。

② 平板电脑 Android 4+。

③ 仿真器 Android 4+。

在仿真器中可以发现其取得的 IP Address 为 10.0.2.15。另外，在 Android 4+ 以后提供支持 IPv6。

仿真器中的相关网络位置如下表所示。

| 网络地址 | 说明 |
| --- | --- |
| 10.0.2.1 | 网关地址 |
| 10.0.2.2 | 原本操作系统的127.0.0.1的别名地址 |
| 10.0.2.3 | 第一优先DNS |
| 10.0.2.4/10.0.2.5/10.0.2.6 | 第二、三、四顺位DNS |
| 10.0.2.15 | AVD本机网络地址 |
| 127.0.0.1 | AVD本机回路地址 |

其中 10.0.2.2 是对应到实际执行 AVD 的操作系统的 127.0.0.1，所以当以 AVD 的观点与 10.0.2.2 的主机进行网络联机时，就是相当于对操作系统的 127.0.0.1 进行网络联机。

## ■ 9-1-3 取得装置联机 IP Address

只要判断网络接口中，非 loopback 的网络接口即为目前对外部联机的 IP Address：

```java
private void getMyIPAddress( ) {
    try {
        for (Enumeration<NetworkInterface> en = NetworkInterface
            .getNetworkInterfaces( ); en.hasMoreElements( );) {
            NetworkInterface intf = en.nextElement( );
            for (Enumeration<InetAddress> enumIpAddr = intf
                    .getInetAddresses( ); enumIpAddr.
                    hasMoreElements( );) {
                InetAddress inetAddress = enumIpAddr.nextElement( );
                if (!inetAddress.isLoopbackAddress( )) {
                    mesg.setText(inetAddress.getHostAddress( ));
                }
            }
        }
    } catch (SocketException ex) {
    }
}
```

### ■ 9-1-4 建构IP Address对象实体

在后面相关因特网单元中，都会应用到 IP Address 对象实体，也就是上例中的 java.net.InetAddress 类别对象实体。该对象实体并非以 new 方式建构出来，通常会通过 InetAddress 类别的 static 方法，调用 getXxx( ) 系列方法取得。例如：

```java
private void textIP( ){
    try {
        InetAddress ip1 = InetAddress.getByName("www.brad.tw");
        InetAddress ip2 = InetAddress.getByName("192.168.12.34");

    } catch (UnknownHostException e) {
        Log.i("brad", "不明主机");
    }
}
```

调用 getByName( ) 方法，传入字符串参数，可以是主机名或是 IP Address。传入主机名之后，会依照用户移动装置的 DNS 进行主机名解析，如果无法解析，则会抛出 UnknownHostException；如果是输入 IP Address 字符串，只要输入格式不正确，也将会抛出 UnknownHostException 的例外异常，但是并非表示该 IP Address 的主机在当时是存在的。

## 9-2 UDP 通信协议的数据存取

UDP 的联机方式是属于非连接导向，当 Client 发送出 UDP 的数据封包给远程 Server 时，如果 Server 已经准备好接收该 UDP 封包数据，当然就马上进行接收；但是当 Server 端并未处在接收状态时，该数据封包就会被自动丢弃，连 Client 端也不会有任何通知重送的相关消息。这样的联机模式重点在于数据传送的速度较快，相对的是并不重视数据的完整性。

### ■ 9-2-1 处理模式

以 java.netDatagramSocket 与 java.net.DatagramPacket 两个类别对象为主。
（1）建立接收端
• 建构 DatagramSocket 对象实体，并绑定在特定的通信端口。
• 建构 DatagramPacket 对象实体，并以特定长度的 byte 数组来接收数据。
• DatagramSocket 对象实体调用 receive( ) 方法，并传入 DatagramPacket 对象实体为参数。
• 接收数据后，相关封包数据内容存放在 DatagramPacket 对象实体中。
而接收完成的 DatagramPacket 对象实体相关数据如下：
• getAddress( )，传回传送端的 InetAddress 对象，可以取得其 IP Address。
• getLength( )，传回传送数据的长度。
• getData( )，传回数据内容 byte 数组。
（2）建立传送端
• 建构 DatagramSocket 对象实体，无须指定通信端口。
• 建构 DatagramPacket 对象实体，将传递的数据内容转换成 byte 数组来传送数据，并指定接收端的 InetAddress 对象，及接收端的通信端口。
• DatagramSocket 对象实体调用 send( ) 方法，并传入 DatagramPacket 对象实体为参数。

### ■ 9-2-2 实作测试

以下直接以 UDP 通信协议进行 Android 移动装置发送数据给一般 PC 电脑。

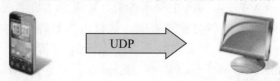

先进行PC接收端，开发以下的Java程序，以线程方式启动UDP等候在特定的通信端口8888，如果读者PC端有防火墙相关设定，记得先将UDP Port:8888先行进行允许通过。

MyUDPReceiver.java

```java
import java.io.IOException;
import java.net.DatagramPacket;
import java.net.DatagramSocket;
import java.net.SocketException;
public class MyUDPReceiver {
    public static void main(String[] args) {
        new UDPReceiveThread( ).start( );
    }
    private static class UDPReceiveThread extends Thread {
        private DatagramSocket socket = null;
        private DatagramPacket packet = null;
        public UDPReceiveThread( ) {
            try {
                socket = new DatagramSocket(8888);
            } catch (SocketException e) {
            }
        }
        @Override
        public void run( ) {
            while (socket != null){
                byte[] buf = new byte[1024];
                packet = new DatagramPacket(buf, buf.length);
                try {
                    socket.receive(packet);
                    String data = new String(packet.getData( ), 0, packet.getLength( ), "UTF-8");
                    System.out.println(data);
                    if (data.equals("quit")){
                        break;
                    }
                } catch (IOException e) {
```

```
                        break;
                }
        }
        if (socket != null) {
                socket.close( );
                socket = null;
        }
    }
  }
}
```

重点说明：

• while 判断 DatagramSocket 是否不为 null，就以一个 1024bytes 的数组为接收数组。

• socket 对象实体调用 receiver( ) 方法传入 packet 对象实体开始等候接收。

• 收到之后调用 packet 对象实体相关方法，取出所需要的数据。

• 如果数据内容为"quit"字符串，则结束等候的线程。

而在 Android 移动装置上开发一个测试项目。规划画面如下：

版面规划：res/layout/activity_main.xml。

```
<LinearLayout xmlns:android="http://schemas.android.com/apk/res/android"
    android:layout_width="match_parent"
    android:layout_height="match_parent"
    android:orientation="vertical" >

    <Button
        android:id="@+id/up"
        android:layout_width="match_parent"
        android:layout_height="wrap_content"
        android:text="Up" />
```

```xml
<LinearLayout
    android:layout_width="match_parent"
    android:layout_height="wrap_content"
    android:orientation="horizontal" >

    <Button
        android:id="@+id/left"
        android:layout_width="match_parent"
        android:layout_height="wrap_content"
        android:layout_weight="1"
        android:text="Left" />

    <Button
        android:id="@+id/stop"
        android:layout_width="match_parent"
        android:layout_height="wrap_content"
        android:layout_weight="1"
        android:text="Stop" />

    <Button
        android:id="@+id/right"
        android:layout_width="match_parent"
        android:layout_height="wrap_content"
        android:layout_weight="1"
        android:text="Right" />

</LinearLayout>

<Button
    android:id="@+id/down"
    android:layout_width="match_parent"
    android:layout_height="wrap_content"
    android:text="Down" />

</LinearLayout>
```

回到程序开发：

```
package tw.brad.android.book.myudpsender;

import java.io.IOException;
```

```java
import java.net.DatagramPacket;
import java.net.DatagramSocket;
import java.net.InetAddress;
import java.net.SocketException;

import android.app.Activity;
import android.os.Bundle;
import android.util.Log;
import android.view.View;
import android.view.View.OnClickListener;

public class MainActivity extends Activity {
    private View up, down, left, right, stop;
    private DatagramSocket socket;

    @Override
    protected void onCreate(Bundle savedInstanceState) {
        super.onCreate(savedInstanceState);
        setContentView(R.layout.activity_main);

        try {
            socket = new DatagramSocket( );
        } catch (SocketException e) {
        }

        up = findViewById(R.id.up);
        down = findViewById(R.id.down);
        left = findViewById(R.id.left);
        right = findViewById(R.id.right);
        stop = findViewById(R.id.stop);

        up.setOnClickListener(new OnClickListener( ) {
            @Override
            public void onClick(View v) {
                new MyUDPSend("往上").start( );
            }
        });
        down.setOnClickListener(new OnClickListener( ) {
            @Override
            public void onClick(View v) {
```

```java
                new MyUDPSend("往下").start( );
            }
        });
        left.setOnClickListener(new OnClickListener( ) {
            @Override
            public void onClick(View v) {
                new MyUDPSend("往左").start( );
            }
        });
        right.setOnClickListener(new OnClickListener( ) {
            @Override
            public void onClick(View v) {
                new MyUDPSend("往右").start( );
            }
        });
        stop.setOnClickListener(new OnClickListener( ) {
            @Override
            public void onClick(View v) {
                new MyUDPSend("quit").start( );
            }
        });
    }
    @Override
    public void finish( ) {
        if (socket != null) socket.close( );
        super.finish( );
    }
    private class MyUDPSend extends Thread {
        private byte data[];
        private DatagramPacket packet;
        MyUDPSend(String dd){
            data = dd.getBytes( );
        }
        @Override
        public void run( ) {
            if (socket != null){
```

```
                try {
                    packet = new DatagramPacket(data, data.length,
                    InetAddress.getByName("192.168.2.104"),
                                    8888);
                        socket.send(packet);
                } catch (IOException e) {
                }
            }
        }
    }
}
```

重点说明：
- 开启权限：INTERNET。
- PC接收端假设其 IP Address 为 192.168.2.104。
- 以上此段测试实作，也可以延伸成为移动装置控制器。

## 9-3　TCP通信协议的数据存取

　　TCP通信协议是属于连接导向，Client端与Server端在进行数据传递之前，必须经过三方握手确认联机状态之后，才开始进行数据的传递。因此，可以确保数据传递的完整性。TCP就好像拨电话一样，一定会确认接通的对方是要找的人之后，才会开始进行信息的传递。

### ■ 9-3-1　处理模式

　　服务器端（Server）先建构出java.net.ServerSocket对象实体即可，传递参数指定其监听在特定的通信部后，调用其accept()方法就开始进行监听（Listening）。一旦客户端（Client）开始进行联机，则将传回一个java.net.Socket对象实体。可以通过Socket对象实体调用getInputStream()传回java.io.InputStream对象实体，而开始进行数据的传递。

　　客户端（Client）只需要建构出java.net.Socket对象实体，指定传送封包数据的目的端及通信部，一旦完成联机，则调用Socket对象实体的getOutputStream()方法传回OutputStream对象实体，开始进行数据传送。

### ■ 9-3-2　实作测试

　　以下直接以TCP通信协议进行Android移动装置发送数据给一般PC电脑。

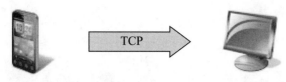

先进行PC服务器端，开发以下的Java程序，以线程方式启动TCP监听在特定的通信端口9999，如果读者PC端有防火墙相关设定，记得先将TCP Port:9999先行进行允许通过。

MyTCPServer.java

```java
import java.io.BufferedReader;
import java.io.IOException;
import java.io.InputStream;
import java.io.InputStreamReader;
import java.net.DatagramPacket;
import java.net.ServerSocket;
import java.net.Socket;
public class MyTCPServer {
    public static void main(String[] args) {
        new TCPServerThread( ).start( );
    }

    private static class TCPServerThread extends Thread {
        private ServerSocket server = null;
        private Socket socket = null;
        private InputStream is = null;
        private BufferedReader reader = null;
        public TCPServerThread( ) {
            try {
                server = new ServerSocket(9999);
            } catch (IOException e) {
            }
        }
        @Override
        public void run( ) {
            stop:
            while (server != null){
                try {
                    socket = server.accept( );
```

```java
                                is = socket.getInputStream( );
                                reader = new BufferedReader(new
                                    InputStreamReader(is));
                                String line;
                                while ( (line = reader.readLine( )) != null){
                                    System.out.println(line);
                                    if (line.equals("quit")) break stop;
                                }
                                reader.close( );
                        } catch (IOException e) {
                        }
                    }
                    if (server != null) {
                        try {
                            server.close( );
                        } catch (IOException e) {
                        }
                    }
                }
            }
}
```

而客户端的版面规划与 UDP 的实作测试一样,仅列出程序开发实作:

```
package tw.brad.android.book.mytcpclient;

import java.io.IOException;
import java.io.OutputStream;
import java.net.InetAddress;
import java.net.Socket;
import java.net.UnknownHostException;

import android.app.Activity;
import android.os.Bundle;
import android.util.Log;
import android.view.View;
import android.view.View.OnClickListener;

public class MainActivity extends Activity {
    private View up, down, left, right, stop;
```

```java
@Override
protected void onCreate(Bundle savedInstanceState) {
    super.onCreate(savedInstanceState);
    setContentView(R.layout.activity_main);

    up = findViewById(R.id.up);
    down = findViewById(R.id.down);
    left = findViewById(R.id.left);
    right = findViewById(R.id.right);
    stop = findViewById(R.id.stop);

    up.setOnClickListener(new OnClickListener( ) {
        @Override
        public void onClick(View v) {
            new MyTCPClient("往上").start( );
        }
    });
    down.setOnClickListener(new OnClickListener( ) {
        @Override
        public void onClick(View v) {
            new MyTCPClient("往下").start( );
        }
    });
    left.setOnClickListener(new OnClickListener( ) {
        @Override
        public void onClick(View v) {
            new MyTCPClient("往左").start( );
        }
    });
    right.setOnClickListener(new OnClickListener( ) {
        @Override
        public void onClick(View v) {
            new MyTCPClient("往右").start( );
        }
    });
    stop.setOnClickListener(new OnClickListener( ) {
        @Override
        public void onClick(View v) {
```

```java
                    new MyTCPClient("quit").start( );
            }
        });
    }

    @Override
    public void finish( ) {
        super.finish( );
    }
    private class MyTCPClient extends Thread {
        private byte data[];
        private Socket socket = null;
        MyTCPClient(String dd) {
            data = dd.getBytes( );
        }

        @Override
        public void run( ) {
            try {
                socket = new Socket(InetAddress.getByName("192.168.2.104"),
                                    9999);
                OutputStream os = socket.getOutputStream( );
                os.write(data);
                os.flush( );
                socket.close( );
            } catch (IOException e) {
                Log.i("brad", "exception");
            }
        }
    }
}
```

重点说明：

• 记得开启权限：INTERNET。

• 所有网络相关对象，必须放在线程中进行，不可以在主线程中进行操作（Android 4+）。

## 9-4 Http通信协议的数据存取

其实Http通信协议是架构在TCP的通信协议，只需要将相关协议的数据传送即可，但是实作上较为复杂，因此在Android上可用较为简便的API来进行开发。

常见的两种API的使用：

• 实作Apache的HttpClient的android.net.http.AndroidHttpClient的类别对象，而在官方API文件是说DefaultHttpClient。

• 以java.net.URL对象实体来开启出HttpURLConnection对象实体。

而以上两个对象实体的概念，就是一般在浏览网页的浏览器的角色地位，所以会直接影响到网页服务器端的Session处理，一旦该对象实体调用close( )方法或是重新取得建构出来，相当于重新开启一个浏览器，此时之前客户端的Session就自动取消。

### ■ 9-4-1　以AndroidHttpClient及DefaultHttpClient实作

而第一种API方式处理，既然与Apache的DefaultHttpClient是类似行为，笔者就先以AndroidHttpClient来实作说明。先来处理GET method方式传递要求。

```
AndroidHttpClient client =
        AndroidHttpClient.newInstance("输入传递给服务器Agent字符串");
```

再来建构出HttpGet对象实体，其概念相当于浏览器上输入的网址：

```
HttpGet get = new HttpGet("http://android.ez2test.com/bombking.png");
```

上例假设想要取得远程服务器的bombking.png的网络资源，事实上是一个png的影像文件。

接着就可以要求client调用执行get对象实体。

```
HttpResponse response = client.execute(get);
```

执行之后会传回一个HttpResponse的对象实体，代表远程服务器的响应对象。

想要取得其响应内容，可以先调用getEntity( )方法传回HttpEntity对象实体，再调用getContent( )方法就可以传回InputStream的输入串流对象，进而接收数据内容。

```
InputStream is = response.getEntity( ).getContent( );
```

以下实作出一个线程来进行这段处理程序如下：

```java
private class MyGetHttp extends Thread {
    String url;
    MyGetHttp(String uu) {
        url = uu;
    }

    @Override
    public void run( ) {
        AndroidHttpClient client = AndroidHttpClient.newInstance("Android");
        HttpGet get = new HttpGet(url);
        try {
            HttpResponse response = client.execute(get);
            InputStream is = response.getEntity( ).getContent( );

            bmp = BitmapFactory.decodeStream(is);
            handler.sendEmptyMessage(1);
        } catch (IOException e) {
            e.printStackTrace( );
        }
    }
}
```

重点说明：
• 声明一个字符串属性用来存放URL。
• 传回来的串流适用于图像文件内容，所以直接调用BitmapFactory.decodeStream( )方法，将数据转为Bitmap对象实体。读者可以转换成其他内容来灵活应用。
• 因为Thread中无法直接进行View组件的存取，所以通过Hander类别的自定义子类别来处理后续的影像呈现在ImageView中。

```java
private class MyHandler extends Handler {
    @Override
    public void handleMessage(Message msg) {
        if (msg.what == 1) {
            img.setImageBitmap(bmp);
        }
    }
}
```

上例中，img 为在版面配置中的 ImageView。

再来以 DefaultHttpClient 来处理 POST method 的传递模式，操作原理非常类似，只是对象实体取得有些许差异，另一方面，既然使用 POST method，也顺便来练习传递参数（GET method 的参数直接以 URL 方式即可）。

建构 DefaultHttpClient 对象实体：

```
DefaultHttpClient client = new DefaultHttpClient( );
```

建构 HttpPost 对象实体：

```
HttpPost post = new HttpPost("http://android.ez2test.com/posttest.php");
```

之后的参数通过 Apache 的 HttpClient 的 API 来进行，因此可以至其官方网站下载（Google 搜索关键词 apache httpclient 即可）。

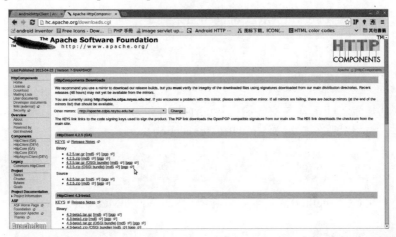

将压缩文件解压至特定目录后，回到 Eclipse 项目下，如果是下载 OSGi bundle 的版本，就是将解压后的一个 org.apache.httpcomponents.httpclient_版本数.jar 文件，或是非 OSGi bundle 的版本，将解压后的 libs/ 子目录下的所有 xxx.jar 文件复制到项目目录下的 libs/ 子目录即可。

开始处理传递的参数，假设有一个需要传递 username 与 password 两个参数，而客户端打算使其值为 brad 与 123456，则先以 org.apache.entity.mime.content.StringBody 类别对象将参数值进行编码。

```
StringBody sb1 = new StringBody("brad");
StringBody sb2 = new StringBody("123456");
```

再来建构出 org.apache.entity.mime.MultipartEntity 对象实体，用来将传递参数建立关系。

```
MultipartEntity entity = new MultipartEntity( );
entity.addPart("username", sb1);
entity.addPart("passwd", sb2);
```

终于要以HttpPost对象实体将其设定为Entity，就可以请DefaultHttp Client对象实体来执行HttpPost了。

```
post.setEntity(entity);
HttpResponse response = client.execute(post);
```

之后的处理动作，从取得InputStream开始都与上述方法一样。

### ■ 9-4-2 以java.net.HttpURLConnection实作

另一种方式是通过HttpURLConnection来处理，通常会先建构出java.net.URL对象实体。

```
URL url = new URL("http://android.ez2test.com/bombking.png");
```

再来就是通过URL对象实体调用其openConnection( )方法，传回URL Connection对象实体，再将其强制转型为HttpURLConnection。

```
HttpURLConnection conn = (HttpURLConnection) url.openConnection( );
```

若是采用GET method的传输，则先设定setRequestMethod( )方法。

```
conn.setRequestMethod("GET");
```

或其他相关设定。

```
conn.setReadTimeout(3000);
conn.setConnectTimeout(5000);
```

最后调用connect( )方法开始进行联机。

```
conn.connect( );
```

没有Exception抛出状况下，就可以接着调用getInputStream( )方法来接收数据。

```
InputStream is = conn.getInputStream( );
```

当取得到InputStream之后，处理的原则和模式就与上述完全相同。

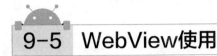

## 9-5 WebView使用

　　android.webkit.WebView是用来呈现网页内容的View组件，其提供了一般浏览器常见的浏览历程（上一页或是下一页）、放大缩小或是文字搜索功能等。网页内容的提供可能是远程的网页服务器，也可能是附加在项目App中的文件。对于网页内容的提供者而言，同时能够符合一般的网页浏览器与移动装置，就不是那么容易。如果将网页内容定位在整个App的版面编排处理，也可以与Android的其他View组件及程序开发控制，这样反而会有不同的开发架构模式。

　　预设状况之下，WebView并非类似浏览器，没有提供支持JavaScript的能力，也就是说定位在以HTML内容来呈现内容为主要目的。如果需要与用户互动的接口，则必须再进一步的设定处理。事实上，如果将与用户互动的层面处理完整的情况下，WebView不失为一个强而有力的用户接口组件。

如果WebView所在的Activity会存取网络上的网页内容，则该Activity必须将INTERNET的权限打开。
专案名称→AndroidManifest.xml中。

```
<uses-permission
android:name="android.permission.INTERNET" />
```

### ■ 9-5-1 基本的处理方式——直接放进Activity中

　　这种方式处理WebView是将Activity调用setContentView( )中设定为一个WebView对象，而该WebView对象设定加载指定的Url字符串的loadUrl( )方法，则将会使其加载指定网址的网页内容呈现整个WebView。

```
import android.os.Bundle;
import android.app.Activity;
import android.view.Menu;
import android.webkit.WebView;
import android.webkit.WebViewClient;
public class MainActivity extends Activity {
```

```
    @Override
    protected void onCreate(Bundle savedInstanceState) {
        super.onCreate(savedInstanceState);
        WebView webView = new WebView(this);
        webView.loadUrl("http://www.pcuser.com.tw/");
        setContentView(webView);
    }
}
```

常见加载网页内容三种方式：
① 直接输入 Http 网址。

```
webView.loadUrl("http://www.pcuser.com.tw/");
```

② 将网页内容文件放在 assets/ 目录下。

```
webview.loadUrl("file:///android_asset/index.html");
```

③ 直接将网页内容以字符串方式加载。

```
String data = "<h1>Brad Big Company</h1><hr />欢迎您光临布莱德大公司";
webview.loadData(data, "text/html; charset=utf-8", null);
```

或

```
String data = "<h1>Brad Big Company</h1><hr />欢迎您光临布莱德大公司";
webview.loadDataWithBaseURL(null, data, null, "utf8", null);
```

### ■ 9-5-2 基本的处理方式——以版面配置方式处理

通常也常见搭配 layout 的 XML 进行配置，增加 Button 组件来处理前后历程，EditText 组件来负责接收用户输入的网址。如果只是将 WebView 组件进行设定 loadUrl( )，则将会相当于以 Intent 方式调用出移动装置用户的默认浏览器进行加载网址网页内容，效果与以下方式类似（不建议使用）：

```
Uri uri = Uri.parse("http://www.pcuser.com.tw");
Intent intent = new Intent(Intent.ACTION_VIEW, uri);
startActivity(intent);
```

因此可以通过WebView组件调用设定setWebViewClient( )方法，传递一个android.webkit.WebViewClient对象即可，也可以自行开发撰写一个WebViewCient的子类别对象，更细部的处理网页内容加载相关程序方法，则WebView组件就可以类似浏览器相关的行为提供给用户。

```
WebViewClient client = new WebViewClient( );
webview.setWebViewClient(client)
```

如下范例，以WebView处理一般浏览器的基本功能。
① 版面配置：

res/layout/activity_main.xml

```xml
<LinearLayout xmlns:android="http://schemas.android.com/apk/res/android"
    android:layout_width="match_parent"
    android:layout_height="match_parent"
    android:orientation="vertical"
    >
    <RelativeLayout
        android:layout_width="match_parent"
        android:layout_height="wrap_content"
    >
        <Button
            android:id="@+id/back"
            android:layout_width="wrap_content"
            android:layout_height="wrap_content"
            android:layout_alignParentLeft="true"
            android:text="向后"
            />
        <Button
            android:id="@+id/forward"
            android:layout_width="wrap_content"
            android:layout_height="wrap_content"
            android:layout_toRightOf="@id/back"
            android:text="向前"
            />
        <Button
            android:id="@+id/reload"
            android:layout_width="wrap_content"
            android:layout_height="wrap_content"
```

```xml
                android:layout_alignParentRight="true"
                android:text="重载"
                    />
    </RelativeLayout>
    <RelativeLayout
        android:layout_width="match_parent"
        android:layout_height="wrap_content"
        >
            <Button
                android:id="@+id/go"
                android:layout_width="wrap_content"
                android:layout_height="wrap_content"
                android:layout_alignParentRight="true"
                android:text="Go"
                    />
    <EditText
        android:id="@+id/url"
            android:layout_width="match_parent"
            android:layout_height="wrap_content"
            android:layout_alignParentLeft="true"
            android:layout_toLeftOf="@id/go"
            android:layout_alignTop="@id/go"
            android:layout_alignBottom="@id/go"
                />
    </RelativeLayout>
    <WebView
        android:id="@+id/webview"
        android:layout_width="match_parent"
        android:layout_height="match_parent"
        />
</LinearLayout>
```

② 程序处理：

```
MainActivity.java package tw.brad.android.books.MyWebViewTest1;

import android.app.Activity;
import android.os.Bundle;
import android.view.View;
import android.view.View.OnClickListener;
import android.webkit.WebView;
import android.webkit.WebViewClient;
```

```java
import android.widget.Button;
import android.widget.EditText;
public class MainActivity extends Activity {
    private Button back, forward, reload, go;
    private EditText url;
    private WebView webview;
    private WebViewClient client;

    @Override
    protected void onCreate(Bundle savedInstanceState) {
        super.onCreate(savedInstanceState);
        setContentView(R.layout.activity_main);

        webview = (WebView)findViewById(R.id.webview);
        url = (EditText)findViewById(R.id.url);
        back = (Button)findViewById(R.id.back);
        forward = (Button)findViewById(R.id.forward);
        reload = (Button)findViewById(R.id.reload);
        go = (Button)findViewById(R.id.go);

        // 设定按下"向后"的按钮功能
        back.setOnClickListener(new OnClickListener( ) {
            @Override
            public void onClick(View v) {
                backWebPage( );
            }
        });

        // 设定按下"向前"的按钮功能
        forward.setOnClickListener(new OnClickListener( ) {
            @Override
            public void onClick(View v) {
                forwardWebPage( );
            }
        });

        // 设定按下"重载"的按钮功能
        reload.setOnClickListener(new OnClickListener( ) {
            @Override
            public void onClick(View v) {
                reloadWebPage( );
            }
        });
```

```java
            // 设定按下"Go"的按钮功能
            go.setOnClickListener(new OnClickListener( ) {
                    @Override
                    public void onClick(View v) {
                            goWebPage( );
                    }
            });
            // 进行 WebView组件初始设定
            initWebView( );
    }
    // 负责WebView初始设定
    private void initWebView( ){
            client = new WebViewClient( );
            webview.setWebViewClient(client);
    }
    // 执行网页浏览历程的上一页
    private void backWebPage( ){
            webview.goBack( );
    }
    // 执行网页浏览历程的下一页
    private void forwardWebPage( ){
            webview.goForward( );
    }
    // 执行网页浏览的重载
    private void reloadWebPage( ){
            webview.reload( );
    }
    // 执行前往指定网页网址
    private void goWebPage( ){
            String gourl = url.getText( ).toString( );
            gourl = "http://" + gourl;
            webview.loadUrl(gourl);
    }
}
```

③ 清单内容：

AndroidManifest.xml

```xml
<?xml version="1.0" encoding="utf-8"?>
```

```xml
<manifest xmlns:android="http://schemas.android.com/apk/res/android"
    package="tw.brad.android.books.MyWebViewTest1"
    android:versionCode="1"
    android:versionName="1.0" >

    <uses-sdk
        android:minSdkVersion="8"
        android:targetSdkVersion="17" />
    <uses-permission android:name="android.permission.INTERNET"/>

    <application
        android:allowBackup="true"
        android:icon="@drawable/ic_launcher"
        android:label="@string/app_name"
        android:theme="@style/AppTheme" >
        <activity
         android:name="tw.brad.android.books.MyWebViewTest1.MainActivity"
            android:label="@string/app_name" >
            <intent-filter>
              <action android:name=" android.intent.action.MAIN" />

                <category android:name="android.intent.category.LAUNCHER" />
            </intent-filter>
        </activity>
    </application>

</manifest>
```

操作使用画面如下图所示。

至于WebViewClient组件的细节处理就是开发撰写其子类别，例如提供当页面开始加载时呈现一个ProgressDialog，而加载完毕后关闭ProgressDialog。

```
//自定义 WebViewClient子类别
private class MyWebClient extends WebViewClient {
    // 当刚开始加载网页内容的事件
    @Override
    public void onPageStarted(WebView view, String url, Bitmap favicon) {
        super.onPageStarted(view, url, favicon);
        progress.show( );
    }
    // 当已经加载网页内容完毕的事件
    @Override
    public void onPageFinished(WebView view, String url) {
        super.onPageFinished(view, url);
        progress.dismiss( );
    }
}
```

就会有以下的效果呈现。

### 9-5-3 进一步设定WebView功能

调用WebView组件的getSettins( )方法传回一个android.webkit.WebSettings对象，经由该对象的设定可以使WebView的功能加强，而其中的JavaScript功能更是后续的延伸使用。

使WebView支持内建缩放尺寸功能：

```
settings = webview.getSettings( );
settings.setBuiltInZoomControls(true);
```

使 WebView 支持 JavaScript 功能：

```
settings.setJavaScriptEnabled(true);
```

使 WebView 不支持密码储存机制（用户安全性考虑）：

```
settings.setSavePassword(false);
```

JavaScript 的作用是在网页内容，WebView 是 Android 中的呈现组件，两者之间的关系就好像是在 PC 上的浏览器观看具有 JavaScript 支持的网页，但是 WebView 可以使这层关系更进化到 JavaScript 可以传递数据给 Android 的 App，而 Android 的 App 其他组件也可以控制到 JavaScript 可以处理的部分，如此一来就可以用网页来开发 Android App 的版面配置及用户互动的层面。

（1）将网页数据传递 Android App

首先来处理简单的网页版面如下（这次的范例将网页文件放在项目下的 assets，读者也可以放在远程网页服务器，加载方式如上文所述）：

```html
asseets/myweb.html
<script>
    function showData( ){
        var data = document.getElementById('data');
        document.getElementById('showhere').innerHTML = data.value;
    }
</script>
<h1>Brad Big Company</h1>
<hr />
<input type="text" id="data" />
<input type="button" onclick="showData( )" value="Click" />
<hr />
<div id="showhere"></div>
```

这段网页的处理机制非常简单，当用户在 id 为 data 中输入文字数据后，按下 Click 的按钮后，将会触发 onclick 事件，进而执行 showData( ) 函数功能，而将输入的文字数据存放在 id 为 showhere 的 div 标签元素中。而这样的动作要能够运行在 Android 中，则必须将 WebSettings 对象的 setJavaScriptEnabled(true) 方法设定，才会生效。

以下解说范例因为网页是放在项目中，也没有进行因特网相关的连接，所以不需要特别去设定INTERNET的权限。

进行版面配置处理：

```
res/layout/activity_main.xml

<LinearLayout xmlns:android="http://schemas.android.com/apk/res/android"
    android:layout_width="match_parent"
    android:layout_height="match_parent"
    android:orientation="vertical"
    >
    <RelativeLayout
        android:layout_width="match_parent"
        android:layout_height="wrap_content"
        >
        <Button
            android:id="@+id/click1"
            android:layout_width="wrap_content"
            android:layout_height="wrap_content"
            android:layout_alignParentLeft="true"
            android:text="Click 1"
            />
        <Button
            android:id="@+id/click2"
            android:layout_width="wrap_content"
            android:layout_height="wrap_content"
            android:layout_toRightOf="@id/click1"
            android:text="Click 2"
            />
    </RelativeLayout>
    <WebView
        android:id="@+id/webview"
        android:layout_width="match_parent"
        android:layout_height="match_parent"
        />
</LinearLayout>
```

两个Button先预留给后面开发使用，目前先来进行WebView的处理。先来最基本的处理设定：

```
MainActivity.java
```

```java
package tw.brad.android.book.MyWebViewTest2;
import android.app.Activity;
import android.os.Bundle;
import android.webkit.WebSettings;
import android.webkit.WebView;
import android.webkit.WebViewClient;
public class MainActivity extends Activity {
    private WebView webview;

    @Override
    protected void onCreate(Bundle savedInstanceState) {
        super.onCreate(savedInstanceState);
        setContentView(R.layout.activity_main);

        webview = (WebView)findViewById(R.id.webview);
        initWebView( );
    }

    private void initWebView( ){
        webview.setWebViewClient(new WebViewClient( ));

        webview.loadUrl("file:///android_asset/myweb.html");
    }
}
```

此时的状况是可以显示出指定的网页内容，但是按下网页中的 Click 按钮却没有任何效果产生，原因就出在没有启用支持 JavaScript 的功能。接下来就改写 initWebView( ) 中的程序如下：

```java
private void initWebView( ){
    webview.setWebViewClient(new WebViewClient( ));

    WebSettings settings = webview.getSettings( );
    settings.setJavaScriptEnabled(true);

    webview.loadUrl("file:///android_asset/myweb.html");
}
```

到此处应该可以正常地看到输入数据后按下 Click 的效果。

但是这只是单纯地在网页中与用户互动，而接下来希望能进一步将用户输入的文字数据传递给 Android App 处理。

先来进行 Android 端的接收机制，只需要撰写非常简单的类别及方法。但是提供给 JavaScript 调用的方法必须是 public 的存取修饰字，切记！

目前放在 MainActivity 类别中的内部类别：

```java
private class MyJavaScript {
    private Context c;

    MyJavaScript(Context c){this.c = c;}
    public void alert(String data){
            Toast.makeText(c, data, Toast.LENGTH_SHORT).show( );
    }
}
```

最主要就是提供了 public void alert( ) 来负责接收 JavaScript 的传递数据。该类别的对象将会被 WebView 对象以调用 addJavascriptInterface( ) 方法传递参数，并设定其字符串名称。

```java
webview.addJavascriptInterface(new MyJavaScript( ), "Brad");
```

以上是指定 MyJavaScript 对象实体的名称为 Brad，该名称将会被 JavaScript 给调用使用。

再回到网页来处理传送：

```
assets/myweb.html 中的 JavaScript。
<script>
    function showData( ){
            var data = document.getElementById('data');
            document.getElementById('showhere').innerHTML = data.value;

            window.Brad.alert(data.value);
            // 或是直接 Brad.alert(data.value);亦可
    }
</script>
```

重点就是调用 Brad 的对象的 alert( ) 方法，就相当于调用 Android 中 MyJavaScript 对象的 alert( ) 方法。

以上的应用面非常广泛，例如以网页版面设计输入窗体，而由 Android App 来处理用户输入的数据；或是呈现 GoogleMap 的 MapView 中，抓到用户点按位置的经纬度数据，而由 Android App 来储存到数据库等。

## （2）由 Android App 传递数据给网页

当用户在其他不是 WebView 组件中，或是存在于程序中的数据，通过 JavaScript 来传递网页进行用户互动。

延续上文的范例，先来处理 JavaScript 的部分，就是撰写一段方法来处理接收自 Android 的数据：

```
assets/myweb.html 中的JavaScript。
<script>
    // 这段函式负责处理接收数据
    function fromAndroid(msg){
        document.getElementById("showhere").innerHTML = msg;
    }
</script>
```

回到 Android 程序中，当要传送给网页调用 JavaScript 函数，一样调用 WebView 对象的 loadUrl( ) 方法传递 "javascript: 函数名称"。

因此修改如下：

① MainActivity.java 中的 onCreate( ) 方法。

```
    click1.setOnClickListener(new OnClickListener( ) {
        @Override
        public void onClick(View v) {
            sendToJavaScript1( );
        }
    });
click2.setOnClickListener(new OnClickListener( ) {
    @Override
    public void onClick(View v) {
        sendToJavaScript2( );
    }
});
```

② MainActivity.java 中对象方法。

```
private void sendToJavaScript1( ){
    webview.loadUrl("javascript:fromAndroid('Hello, World')");
}
private void sendToJavaScript2( ){
```

```
    webview.loadUrl("javascript:fromAndroid('" + (int)(Math.
random( )*49+1)+ "')");
}
```

分别按下Click1按钮会出现"Hello，World"；按下Click2会出现1～49之间的随机数。

可以应用在移动装置的数据传递到网页内容来处理，例如将用户目前GPS所侦测到的地理位置传给网页。

# 10 Chapter

## 第10课　影音多媒体与相机

10-1　播放音乐

10-2　音效处理

10-3　录音处理

10-4　录像放映

10-5　相机

## 10-1 播放音乐

### ■ 10-1-1 基本概念

在移动装置上面播放音乐应该算是使用频率相当高的项目。而在一般应用程序的App中可能没有那么常用，除非该应用程序就是以播放音乐为主，但是在游戏App中几乎是不可或缺的功能。

播放音乐的类别主要是通过android.media.MediaPlayer类别对象实体来进行，通常会有以下几种播放状况：

- 播放SDCard上面的音乐文件。
- 播放项目资源中的音乐文件（例如游戏中的背景音乐）。
- 播放因特网远程的音乐文件。

MediaPlayer对象实体播放音乐的状态图如下所示。

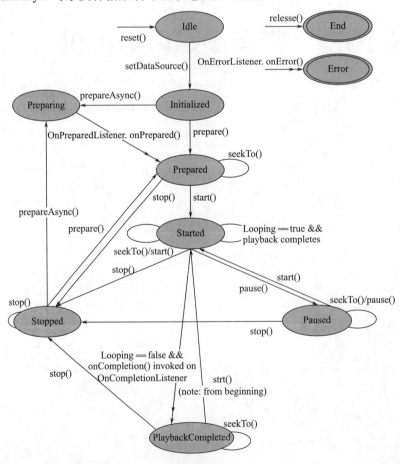

重点说明：
- 图中的椭圆形表示状态。
- 调用 setDataSource( ) 会使其进入初始完成状态。
- 完成初始状态继续调用 prepare( ) 进入完成准备状态。
- 开始播放音乐是从 start( ) 开始。
- 调用 stop( ) 会进入停止状态。
- 播放状态中调用 pause( ) 会进入暂停状态。
- 暂停状态下调用 start( ) 后又继续开始播放。
- 播放及暂停状态下可以调用 seekTo( ) 进行跳跃。
- 调用 reset( ) 会使其进入闲置状态。
- 调用 release( ) 会使其进入结束状态。

## ■ 10-1-2　SDCard上的音乐播放

先假设要播放的音乐文件放在 SDCard 下的 brad.mp3，所以在程序中：

```
Environment.getExternalStorageDirectory( ).getPath( ) +
                                          "/brad.mp3"
```

开始建构 MediaPlayer 对象实体。

```
player = new MediaPlayer( );
```

并设定其数据来源。

```
player.setDataSource(Environment.getExternalStorageDirectory( )
.getPath( ) + "/bg01.mp3");
```

使其进入准备状态后直接播放。

```
try {
    player.setDataSource(Environment
        .getExternalStorageDirectory( )
        .getPath( ) + "/brad.mp3");
    player.prepare( );
    player.start( );
} catch (IOException e) {
    e.printStackTrace( );
}
```

其他常见的控制方法调用：
- setVolume (左边喇叭音量 float 值, 右边喇叭音量 float 值)：设定播放音量。
- getDuration( )：传回目前播放时间（1/1000s 单位）。
- seekTo（跳跃到的时间 1/1000s 单位）。

可以搭配显示播放进度，利用 SeekBar 类别对象实体来实作。先在 MediaPlayer 对象实体调用 start( ) 方法之后，开始将目前的 Duration 值设定给 SeekBar 对象实体。

```
seekbar.setMax(player.getDuration( ));
```

并且通过 Timer/TimerTask 来周期将目前播放的进度传回。

```
private class SeekBarTask extends TimerTask {
    @Override
    public void run( ) {
        seekbar.setProgress(player.getCurrentPosition( ));
    }
}
```

而在一开始播放时启动周期任务。

```
SBTask = new SeekBarTask( );
timer.scheduleAtFixedRate(SBTask, 0, 200);
```

接着在 SeekBar 对象实体处理 setOnSeekBarChangeListener( )。

```
seekbar.setOnSeekBarChangeListener(new OnSeekBarChangeListener( ) {
    @Override
    public void onStopTrackingTouch(SeekBar seekBar) {
    }
    @Override
    public void onStartTrackingTouch(SeekBar seekBar) {
    }
    @Override
    public void onProgressChanged(SeekBar seekBar, int progress,
            boolean fromUser) {
        if (fromUser && player != null) {
            player.seekTo(progress);
```

```
            } else if (!fromUser) {
                info.setText(progress + "/" + seekbar.getMax( ));
            }
        }
});
```

### ■ 10-1-3　播放项目资源中音乐文件

先将要播放的音乐文件放进项目下 res/raw/ 子目录中。假设放了一个 res/raw/brad.mp3，这种类型的播放并不需要以 new 方式建构出 MediaPlayer 对象实体，而是调用 MediaPlayer 类别的 static 方法 create( )，将会传回指定资源的 MediaPlayer 对象实体。

```
player = MediaPlayer.create(this, R.raw.brad);
```

如果是游戏的背景音乐，通常会先调用 setLooping(true)，使音乐会不断地重复播放。当然这种状况下应该去设计的音乐结尾与开头会有连续性，才不会使玩家感觉明显的开始与结束。

```
player.setLooping(true);
player.start( );
```

### ■ 10-1-4　播放URL的音乐文件

播放模式与SDCard相同，但是因为远程音乐文件相较于SDCard上面的文件，可能会需要较多的准备工作，因此不是调用 prepare( ) 方法，而是非同步的 prepareAsync( ) 方法。而要等到准备完成之后才能开始 start( )，但是何时会准备好呢？可以通过 MediaPlayer 对象实体设定 setOnPreparedListener( ) 来进行监听。

```
try {
    player.setDataSource("http://www.ez2test.com/brad.mp3");
    // player.prepare( );
    player.prepareAsync( );
    player.setOnPreparedListener(new OnPreparedListener( ) {
        @Override
        public void onPrepared(MediaPlayer mp) {
            player.start( );
```

```
                    seekbar.setMax(player.getDuration( ));
                    SBTask = new SeekBarTask( );
                    timer.scheduleAtFixedRate(SBTask, 0, 200);
            }
        });
} catch (IOException e) {
    e.printStackTrace( );
}
```

 别忘记要开启 INTERNET 的使用权限。

支持的通信协议如下：
① RTSP (RTP，SDP)。
② HTTP/HTTPS progressive streaming。
③ HTTP/HTTPS live streaming：
• MPEG-2 TS media files only。
• Protocol version 3 (Android 4.0 and above)。
• Protocol version 2 (Android 3.x)。
• Not supported before Android 3.0。

### ■ 10-1-5　暂停继续播放

应该会先调用 MediaPlayer 对象实体的 isPlaying( ) 方法传回 boolean 值判断目前是否播放状态中。

```
if (player.isPlaying( )) {
    // 当作暂停处理
    player.pause( );
} else if (!player.isPlaying( )) {
    // 当作继续处理
    player.start( );
}
```

### ■ 10-1-6　停止播放

只是调用 stop( ) 方法，但是如果是应用程序结束的话，则应该调用 reset( )。另外，如果有使用到 Timer/TimerTask 来处理 SeekBar 对象，也别忘记将 Timer 对象实体取消结束。

```
public void finish( ) {
    if (player != null && player.isPlaying( )) {
          player.reset( );
          player = null;
    }
    timer.cancel( );
    timer = null;
    super.finish( );
}
private void stopMusic( ) {
    if (player != null) {
          player.stop( );
          player = null;
          SBTask.cancel( );
          seekbar.setProgress(0);
    }
}
```

## 10-2 音效处理

音效与音乐不同之处在于音效有较高的实时性，当发射子弹后就要马上发出"咻"的音效，稍微延迟就不合游戏情境。因此处理的观念就是事先载入，当要播放音效时直接调用，如果还要做准备工作就太迟了。通常会以 SoundPool 来实作。

### ■ 10-2-1 建构SoundPool对象实体

```
spool = new SoundPool(10, AudioManager.STREAM_MUSIC, 0);
```

传递以下参数项：
- 最大串流数。
- 串流类型。
- 取样转换值，默认值为 0，目前尚无效果。

将要播放的文件放在 res/raw/ 子目录下。先定义播放音效。

```
private int[] sounds = new int[4];
private static final int SOUND_HIT = 0;
private static final int SOUND_GO = 1;
private static final int SOUND_KILL = 2;
private static final int SOUND_OK = 3;
```

开始加载声音文件,调用load( )方法,传回特定的整数音效ID。

```
sounds[SOUND_HIT] = spool.load(this, R.raw.bang_1, 1);
sounds[SOUND_GO] = spool.load(this, R.raw.bang_3, 1);
sounds[SOUND_KILL] = spool.load(this, R.raw.cling_steel, 1);
sounds[SOUND_OK] = spool.load(this, R.raw.freebar, 1);
```

### ■ 10-2-2　实时播放音效

调用SoundPool对象实体的play( )方法,传递以下参数:
- 事先加载的音效ID。
- 左声道音量。
- 右声道音量。
- 优先级。
- 是否重复。
- 取样值。

```
spool.play(sounds[SOUND_HIT], rvol * 0.1f, rvol * 0.1f, 1, 0, 1);
```

## 10-3　录音处理

通过移动装置的麦克风可以进行录音工作。处理模式大致上有两种模式:
- 通过Intent调用其他录音程序。
- 自定义处理录音程序。

录音工作有两个权限必须开启(Uses Permission):
- 录音:RECORD_AUDIO。
- 写档:WRITE_EXTERNAL_STORAGE。

```
<uses-permission android:name="android.permission.RECORD_AUDIO"/>
<uses-permission android:name="android.permission.WRITE_EXTERNAL_STORAGE"/>
```

## 10-3-1 调用其他录音程序

这种模式处理比较简单,只需要通过 Intent 传递调用即可。

```
Intent intent =
    new Intent(MediaStore.Audio.Media.RECORD_SOUND_ACTION);
startActivityForResult(intent, 0);
```

在不同的移动装置上的不同调用:

录音完毕之后的存盘对话框:

而录音完毕之后，会传回一个Recult Code给原本调用的Activity，因此，原本调用的Activity必须要改写onActivityResult( )才能接收其值。录音存盘完毕，将会回传值为RESULT_OK(=-1)。并将其数据放在Intent对象实体中，通过调用getData( )传回android.net.Uri的对象实体，如果想要进一步取得实体文件，则必须另外开发撰写转换方法。

```java
@Override
protected void onActivityResult(int requestCode, int resultCode, Intent data) {
    Log.i("brad", "Result: " + resultCode);
    if (resultCode == RESULT_OK){
        Uri uri = data.getData( );
        Log.i("brad", uri.getPath( ));
        Log.i("brad", getRealPathFromURI(uri));
        realFile = getRealPathFromURI(uri);
    }else {
        Log.i("brad", "Cancel Record");
    }
}
```

附上一段自定义的转换方法：

```java
public String getRealPathFromURI(Uri contentUri) {
    String[] proj = { MediaStore.Audio.Media.DATA };
    Cursor cursor = getContentResolver( ).query(contentUri, proj, null, null, null);
    int column_index = cursor
            .getColumnIndexOrThrow(MediaStore.Audio.Media.DATA);
    cursor.moveToFirst( );
    return cursor.getString(column_index);
}
```

### ■ 10-3-2　自定义录音处理程序

自定义录音处理程序就略为复杂些，要把握住录音的状态控制即可。整个操作的主角为android.media.MediaRecorder的对象实体，以下先了解其状态程序图。

录音状态图如下：

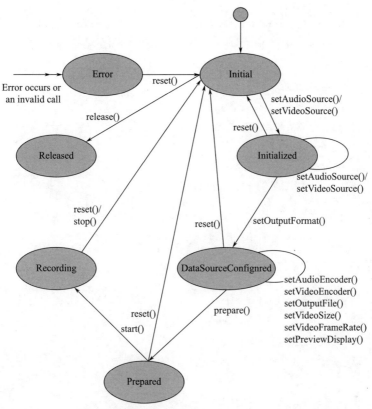

MediaRecorder state diagram

重点说明:
- 进行准备之前的调用程序的顺序性非常重要。
- 先调用 setAudioSource( ) 完成初始化状态。
- 再调用 setOutputFormat( ) 设定输出格式,完成数据源组态。
- 再来设定 setAudioEncoder( ) 编码器,及 setOuputFile( ) 输出文件。
- 调用 prepare( ) 进入准备完成状态。
- 终于可以开始调用 start( ) 进行录音。
- 录音期间调用 stop( ) 则停止录音。

建构 MediaRecorder 对象实体:

```
recorder = new MediaRecorder( );
```

开始准播程序及录音:

```
private void startRecorder( ) {
    recorder.setAudioSource(MediaRecorder.AudioSource.MIC);
    recorder.setOutputFormat(MediaRecorder.OutputFormat.MPEG_4);
```

```
        recorder.setAudioEncoder(MediaRecorder.AudioEncoder.DEFAULT);
        recorder.setOutputFile("/mnt/sdcard/bradrecorder.mp4");
        try {
                recorder.prepare( );
                recorder.start( );
        } catch (IllegalStateException e) {
                e.printStackTrace( );
        } catch (IOException e) {
                e.printStackTrace( );
        }
}
```

停止录音:

```
private void stopRecorder( ) {
    try {
            recorder.stop( );
    } catch (IllegalStateException ee) {
    }
}
```

## 10-4 录像放映

### ■ 10-4-1 录像

录像的处理模式与录音非常类似，也是有两种方式可以处理。
- 调用其他录像程序。
- 自定义录像程序。

录像的处理有三个权限需要开启：
- 相机：CAMERA（调用其他录像程序时不需要）。
- 录音：RECORD_AUDIO（调用其他录像程序时不需要）。
- 写档：WRITE_EXTERNAL_STORAGE（两种处理模式都需要）。

### ■ 10-4-2 调用其他录像程序

原理与录音类似，但是可以事先设定录像的文件，这样就可以不必在事后进行

Uri 的实体文件名转换。

先来了解与录音类似的手法。

```
Intent intent = new Intent(
        android.provider.MediaStore.ACTION_VIDEO_CAPTURE);
startActivityForResult(intent, 1);
```

改写 onActivityResult( )方法：

```
@Override
protected void onActivityResult(int requestCode, int resultCode, Intent data) {
    if (resultCode == RESULT_OK && requestCode == 1){
        Uri uri = data.getData( );
        Log.i("brad", uri.getPath( ));
        Log.i("brad", getRealPathFromURI(uri));
    }
}

public String getRealPathFromURI(Uri contentUri) {
    String[] proj = { MediaStore.Audio.Media.DATA };
    Cursor cursor = getContentResolver( ).query(contentUri, proj, null, null, null);
    int column_index = cursor
            .getColumnIndexOrThrow(MediaStore.Audio.Media.DATA);
    cursor.moveToFirst( );
    return cursor.getString(column_index);
}
```

也可以事先设定录像文件，如下处理模式：

```
private void doVR2( ) {
    Intent intent = new Intent(
            android.provider.MediaStore.ACTION_VIDEO_CAPTURE);
    intent.putExtra(MediaStore.EXTRA_OUTPUT, Uri.fromFile(
new File("/mnt/sdcard/ok.3gp")));
    startActivityForResult(intent, 2);
}
```

## 10-4-3 自定义录像程序

通过相机装置进行录像，此时要使用SurfaceView来处理屏幕变动更新较为频繁的状态。也是与录音相同地使用MediaRecorder对象实体进行。

先来设计录像的视景，在版面配置文件中，以SurfaceView来处理。

```xml
<SurfaceView
    android:id="@+id/sv"
    android:layout_width="match_parent"
    android:layout_height="match_parent" />
```

进行重要对象变量声明：

```java
private SurfaceView sv;
private SurfaceHolder holder;
private Camera camera;
private MediaRecorder mr;
```

进行建构对象实体及设定：

```java
SetRequestedOrientation(
ActivityInfo.SCREEN_ORIENTATION_LANDSCAPE);
sv = (SurfaceView) findViewById(R.id.sv);
holder = sv.getHolder( );
holder.addCallback(this);
holder.setType(SurfaceHolder.SURFACE_TYPE_PUSH_BUFFERS);
```

此时因为addCallback(this)，所以Activity必须声明implements android.view.SurfaceHolder.Callback，并且实作以下方法：

```java
@Override
public void surfaceChanged(SurfaceHolder holder, int format, int width,
    int height) {
}

@Override
public void surfaceCreated(SurfaceHolder holder) {
}
```

```java
@Override
public void surfaceDestroyed(SurfaceHolder holder) {

}
```

继续针对相机及 **MediaRecorder** 对象进行初始设定。

```java
private void initRecorder(Surface surface) {
    if (camera == null) {
        camera = Camera.open( );
        camera.unlock( );
    }

    if (mr == null) {
        mr = new MediaRecorder( );
    }

    mr.setPreviewDisplay(surface);
    mr.setCamera(camera);

    mr.setVideoSource(MediaRecorder.VideoSource.CAMERA);
    mr.setOutputFormat(MediaRecorder.OutputFormat.DEFAULT);
    mr.setOutputFile("/mnt/sdcard/test.mp4");
    mr.setMaxDuration(-1);  // 没有限制录像时限
    mr.setVideoFrameRate(15);
    mr.setVideoEncoder(MediaRecorder.VideoEncoder.DEFAULT);
    try {
        mr.prepare( );
    } catch (IllegalStateException e) {
        // TODO Auto-generated catch block
        e.printStackTrace( );
    } catch (IOException e) {
        // TODO Auto-generated catch block
        e.printStackTrace( );
    }
}
```

假设设计一个 **Button** 第一次按下去开始录像，第二次按下去结束录像：

```java
private void doVR( ) {
    if (isRecording) {
```

```
            mr.stop( );
            mr.reset( );
    } else {
            initRecorder(holder.getSurface( ));
            mr.start( );
    }
    isRecording = !isRecording;
}
```

最后很重要的处理动作，就是针对App结束之后的释放装置，否则用户很可能离开程序之后就无法使用相机，直到下次重新开启移动装置。

```
@Override
public void surfaceDestroyed(SurfaceHolder holder) {
    if (mr != null) {
            mr.reset( );
            mr.release( );
            mr = null;
    if (camera != null) {
            camera.release( );
            camera = null;
    }
}
```

### ■ 10-4-4 播放影片

可以通过VideoView来进行影片播放，先来处理版面配置。

```
<RelativeLayout xmlns:android="http://schemas.android.com/apk/res/android"
    android:layout_width="match_parent"
    android:layout_height="match_parent"
    >
    <VideoView
        android:id="@+id/vv"
        android:layout_width="match_parent"
        android:layout_height="match_parent"
        />
</RelativeLayout>
```

操作方式更为简单，找出VideoView对象实体的参考。

```
vv = (VideoView)findViewById(R.id.vv);
```

播放SDCard：

```
vv.setVideoPath("/mnt/sdcard/test.mp4");
```

播放res/raw/目录下的资源：

```
vv.setVideoURI(Uri.parse("android.resource://" + getPackageName( ) + "/" + R.raw.test));
```

播放远程文件：

```
vv.setVideoURI(Uri .parse("http://www.ez2test.com/brad.mp4"));
```

开始播放：

```
vv.requestFocus( );
vv.start( );
```

## 10-5 相机

相机拍照的处理方式，也可以和录音录像一样，也是有两种方式处理。
- 调用其他照相程序。
- 自定义照相程序。

照相的处理有两个权限必须开启：
- 相机：CAMERA（调用其他照相程序时不需要）。
- 写档：WRITE_EXTERNAL_STORAGE（两种处理模式都需要）。

### 10-5-1 调用其他照相程序

处理模式与录像类似，但是因为是照相处理，笔者再区分出缩图和原图，因为往往在许多应用上，会使用到照相之后的缩图，而不需要另外在程序中处理。而缩图与原图也可以同时一次照相取得，只是为了讲解方便而区分说明。

直接以Intent对象实体交给系统来调用用户已经安装的照相程序菜单。

```
private void takePic1( ){
    Intent intent = new Intent(MediaStore.ACTION_IMAGE_CAPTURE);
    startActivityForResult(intent, 1);
}
```

如下图所示。

 如果用户之前已经设定了预设相机，或是只有一个相机程式，那就不会出现如上图的对话框。

当照相完成并在该照相程序中确定后，则会回传ResultCode为RESULT_OK。

```
protected void onActivityResult(int requestCode, int resultCode, Intent data)
    { super.onActivityResult(requestCode, resultCode, data);
    if (resultCode == RESULT_OK){
        switch (requestCode){
            case 1:
                afterTakePic1(data);
                break;
            case 2:
                afterTakePic2( );
                break;
            case 3:
                afterTakePic3( );
```

```
                    break;
            }
        }
}
```

以下分别处理 afterTakePicX( )。

只取得缩图 afterTakePic1( )：

```
private void afterTakePic1(Intent data){
    Bitmap bmp = (Bitmap)data.getExtras( ).get("data");    // 缩图
    img.setImageBitmap(bmp);

    // 如果喜欢的话，将该缩图另存文件
    try {
        bmp.compress(CompressFormat.JPEG, 85, new FileOutputStream("/
            mnt/sdcard/camera1.jpg"));
    } catch (FileNotFoundException e) {
    }
}
```

若要直接在照相之前，就先设定照相原文件的放置，则会在Intent对象实体中设定。

```
private void takePic2( ){
    Intent intent = new Intent(MediaStore.ACTION_IMAGE_CAPTURE);
    outputFileUri = Uri.fromFile(new File("/mnt/sdcard/camera2.jpg"));
    intent.putExtra(MediaStore.EXTRA_OUTPUT, outputFileUri);
    startActivityForResult(intent, 2);
}
```

接着处理 afterTakePic2( )：

```
private void afterTakePic2( ){
    Bitmap bmp = BitmapFactory.decodeFile("/mnt/sdcard/camera2.jpg");
    img.setImageBitmap(bmp);
}
```

### ■ 10-5-2　自定义相机程序

当然一切就要从头处理，所以就先从是否有相机装置，有几个相机装置开始下手。

在Activity中调用getPackageManager( )方法传回PackageManager对象实体，再由PackageManager对象实体调用其hasSystemFeature( )方法，传递PackageManager.FEATURE_CAMERA参数，传回boolean值表示是否有相机装置。其实，不妨借此机会在这里稍微了解更多的装置侦测。例如许多读者认为用户的装置为API Level 9+，所以就可以使用NFC，这是不正确的观念。API是软件的支持，而硬件装置是否有配置是两回事，所以要通过侦测方式来判断。

```
private void takePic3( ) {
    int numCamera = checkCameraHardware(this);
    msg.setText("Camera num:" + numCamera);
    if (numCamera > 0){
    }
}

// 侦测相机硬件
private int checkCameraHardware(Context context) {
    PackageManager pkgmgr = getPackageManager( );
    if (pkgmgr.hasSystemFeature(PackageManager.FEATURE_CAMERA)){
        // 有相机装置
      return Camera.getNumberOfCameras( );// 传回相机装置个数
      API Level 9+
    } else {
        // 无相机装置
        return 0;
    }
}
```

可以继续针对相机装置来取得其支持能力。

```
// 取得相机对象实体
public static Camera getCameraInstance(int cid){
    Camera c = null;
    try {
        c = Camera.open(cid);
    }
    catch (Exception e){
    return c;
}
```

```
// 侦测相机支持功能
private void checkCameraSupport( ){
    Camera c = getCameraInstance(0);
    Camera.Parameters param = c.getParameters( );
    c.release( );
}
```

开始进行照相的动作。

自定义预览观景窗是自定义相机要先处理的部分,这是一个继承android.view.SurfaceView的自定义类别,并且实作android.view.SurfaceHolder.Callback界面。

CameraPreview.java

```
package tw.brad.android.book.mycamera;
import java.io.IOException;
import android.content.Context;
import android.hardware.Camera;
import android.util.Log;
import android.view.SurfaceHolder;
import android.view.SurfaceView;
public class CameraPreview extends SurfaceView implements SurfaceHolder.Callback {
    private SurfaceHolder mHolder;
    private Camera mCamera;
    public CameraPreview(Context context, Camera camera) {
        super(context);
        mCamera = camera;

        mHolder = getHolder( );
        mHolder.addCallback(this);
        mHolder.setType(SurfaceHolder.SURFACE_TYPE_PUSH_BUFFERS);
    }
    public void surfaceCreated(SurfaceHolder holder) {
        try {
            mCamera.setPreviewDisplay(holder);
            mCamera.startPreview( );
        } catch (IOException e) {
```

```
            Log.d("brad","预览观景窗设定错误");
        }
    }
    public void surfaceDestroyed(SurfaceHolder holder) {
        // 结束观景窗后，也将相机实体释放
     mCamera.release( );
     mCamera = null;
    } public void surfaceChanged(SurfaceHolder holder, int format,
int w, int h){
        if (mHolder.getSurface( ) == null){
           return;
        try {
            mCamera.stopPreview( );
        } catch (Exception e){
        }
        try {
            mCamera.setPreviewDisplay(mHolder);
            mCamera.startPreview( );
        } catch (Exception e){
            Log.d("brad", "错误设定观景窗");
        }
    }
}
```

接着来处理整体的预览版面配置。

res/layout/activity_brad_camera.xml

```
<LinearLayout xmlns:android="http://schemas.android.com/apk/res/android"
    android:layout_width="fill_parent"
    android:layout_height="fill_parent"
    android:orientation="horizontal" >

    <FrameLayout
        android:id="@+id/camera_preview"
        android:layout_width="fill_parent"
        android:layout_height="fill_parent"
        android:layout_weight="1" />
```

```xml
    <Button
        android:id="@+id/button_capture"
        android:layout_width="wrap_content"
        android:layout_height="wrap_content"
        android:layout_gravity="center"
        android:text="Capture" />
</LinearLayout>
```

Button 是用来触发照相的动作，而 FrameLayout 是用来放置刚刚处理好的 CameraPreview 用的。

终于可以来写照相的程序。

BradCamera.java

```java
package tw.brad.android.book.mycamera;

import java.io.File;
import java.io.FileNotFoundException;
import java.io.FileOutputStream;
import java.io.IOException;

import android.app.Activity;
import android.hardware.Camera;
import android.hardware.Camera.PictureCallback;
import android.os.Bundle;
import android.util.Log;
import android.view.View;
import android.view.View.OnClickListener;
import android.widget.Button;
import android.widget.FrameLayout;

public class BradCamera extends Activity {
    private Camera mCamera;
    private CameraPreview mPreview;
    private Button capture;

    @Override
    public void onCreate(Bundle savedInstanceState) {
        super.onCreate(savedInstanceState);
        setContentView(R.layout.activity_brad_camera);

        mCamera = getCameraInstance(0);
```

```java
            mPreview = new CameraPreview(this, mCamera);
            FrameLayout preview = (FrameLayout) findViewById(R.id.camera_preview);
            preview.addView(mPreview);

            capture = (Button)findViewById(R.id.button_capture);
            capture.setOnClickListener(new OnClickListener( ) {
                @Override
                public void onClick(View v) {
                    mCamera.takePicture(null, null, new MyPicCallback( ));
                }
            });
    }

    private class MyPicCallback implements PictureCallback {
        @Override
        public void onPictureTaken(byte[] data, Camera camera) {
          File pictureFile = new File("/mnt/sdcard/camera3.jpg");
          if (pictureFile == null){
              Log.d("brad", "文件写出权限失败");
              return;
           } try {

            FileOutputStream fos = new FileOutputStream(pictureFile);
              fos.write(data);
              fos.close( );
           } catch (FileNotFoundException e) {
           } catch (IOException e) {
           }
         }
    }

    // 取得相机对象实体
    public static Camera getCameraInstance(int cid) {
        Camera c = null;
        try {
                c = Camera.open(cid);
        } catch (Exception e) {
        }
        return c;
    }
}
```

# 第11课 地图与卫星定位系统

11-1　GPS定位

11-2　基本Google Map

11-3　进阶Google Map

## 11-1　GPS定位

通过Android移动装置的卫星定位系统，可以取得目前所在的位置信息，通常就是使用经纬度来表示出定位信息。定位信息的取得方式通常有两种：GPS及网络。GPS仅适用户外的环境下，因此使用上大受限制；而使用移动网络方式定位，相对不够精确，优点就是耗电少。如下表所示。

| 比较项目/定位方式 | GPS | Network |
| --- | --- | --- |
| 精确度 | 高 | 差 |
| 耗电量 | 高 | 低 |
| 高度(海拔) | 有 | 无 |
| 速度 | 有 | 无 |

通过取得的经纬度信息，搭配使用Google Map，可以让用户轻松地了解目前所在位置的相关地理信息。常见的应用就是将所在位置附近的商家或是旅游景点提供详细数据，或是将定位路线产生路径图的呈现等。

开启用户权限如下：
- GPS定位：ACCESS_FINE_LOCATION。
- Network定位：ACCESS_COARSE_LOCATION。

### ■ 11-1-1　开始基本实作

声明使用LocationManager类别对象，该对象是整个定位系统的主要管理对象。

```
private LocationManager lmgr;
```

通过调用getSystemService( )方法，传入LOCATION_SERVICE参数，将传回值强制转型为LocationManager。

```
lmgr = (LocationManager)getSystemService(LOCATION_SERVICE);
```

而定位的提供者有两个变量来处理：

```
locationProvider = LocationManager.NETWORK_PROVIDER;
locationProvider = LocationManager.GPS_PROVIDER;
```

接着就开始处理定位监听对象，只需要自定义类别implements android.location.LocationListener界面即可：

```java
private class MyLocationListener implements LocationListener {
    @Override
    public void onLocationChanged(Location location) {
    }
    @Override
    public void onProviderDisabled(String provider) {
    }
    @Override
    public void onProviderEnabled(String provider) {
    }
    @Override
    public void onStatusChanged(String provider, int status, Bundle extras) {
    }
}
```

其中最主要实作的方法就是 onLocationChanged( )，该方法会在用户的移动装置的定位位置移动改变，而触发传入的参数 Location 对象实体，涵盖了目前位置的相关信息，常见如下数据：

- getAccuracy( )：传回精准度，其数值单位为公尺（float）。
- getAltitude( )：传回海拔高度，其数值单位为公尺（double）。
- getBearing( )：传回方位角度，其数值单位为角度 0.0～360.0（float）。
- getLatitude( )：传回纬度，其数值单位为度（double）。
- getLongitude( )：传回经度，其数值单位为角度（double）。
- getSpeed( )：传回速度，其数值单位为公尺/秒（float）。
- getTime( )：传回时间值，表示从 1970 年 1 月 1 日起算的 1/1000s。

最后就通过 LocationManager 对象实体调用 requestLocationUpdates( ) 方法，传入以下参数：

- 定位提供者。
- 位移更新最小时间，单位为 1/1000s。
- 位移更新最小距离，单位为公尺。
- 定位监听对象实体。

```java
mll = new MyLocationListener( );
lmgr.requestLocationUpdates(locationProvider, 0, 0, mll);
```

当不再需要使用 GPS 定位时，记得将该定位监听对象解除注册。

```
lmgr.removeUpdates(mll);
```

## ■ 11-1-2 较佳位置取得

此段内容参考 Google Android 开发者网站的文件,试图在 GPS 与 Network 的定位上进行切换,取得较佳的位置信息内容。

建立一个自定义类别,用来处理用户的位置对象相关属性及方法。该类别定义了几个重要的属性:

- 排程定时器对象 Timer,用来周期处理定位程序。
- 定位管理员 LocationManager 对象。
- 自定义定位结果对象 LocationResult。

如下简单的程序代码 MyLocation.java。

```java
package tw.brad.android.book.mygpsandmapv3;

import java.util.Timer;
import java.util.TimerTask;
import android.content.Context;
import android.location.Location;
import android.location.LocationListener;
import android.location.LocationManager;
import android.os.Bundle;

public class MyLocation {
    Timer timer1;
    LocationManager lm;
    LocationResult locationResult;
    boolean gps_enabled = false;
    boolean network_enabled = false;

    public boolean getLocation(Context context, LocationResult result) {
            locationResult = result;
            if (lm == null)
                    lm = (LocationManager) context
                        .getSystemService(Context.LOCATION_SERVICE);
            try {
             gps_enabled = lm.isProviderEnabled(LocationManager.GPS_
             PROVIDER);
```

```
        } catch (Exception ex) {
        }
        try {
              network_enabled = lm
.isProviderEnabled(LocationManager.NETWORK_PROVIDER);
        } catch (Exception ex) {
        }

        if (!gps_enabled && !network_enabled)
              return false;

        if (gps_enabled)
        lm.requestLocationUpdates(LocationManager.GPS_PROVIDER,
                  0, 0, locationListenerGps);
        if (network_enabled)
lm.requestLocationUpdates(LocationManager.NETWORK_PROVIDER, 0, 0,
              locationListenerNetwork);
        timer1 = new Timer( );
        timer1.schedule(new GetLastLocation( ), 20000);
        return true;
}

LocationListener locationListenerGps = new LocationListener( ) {
        public void onLocationChanged(Location location) {
              timer1.cancel( );
              locationResult.gotLocation(location);
              lm.removeUpdates(this);
              lm.removeUpdates(locationListenerNetwork);
        }
        public void onProviderDisabled(String provider) {
        }
        public void onProviderEnabled(String provider) {
        }
        public void onStatusChanged(String provider, int status,
        Bundle extras) {
        }
};
```

```java
LocationListener locationListenerNetwork = new LocationListener( ) {
    public void onLocationChanged(Location location) {
        timer1.cancel( );
        locationResult.gotLocation(location);
        lm.removeUpdates(this);
        lm.removeUpdates(locationListenerGps);
    }

    public void onProviderDisabled(String provider) {
    }

    public void onProviderEnabled(String provider) {
    }
    public void onStatusChanged(String provider, int status, Bundle extras) {
    }
};

class GetLastLocation extends TimerTask {
    @Override
    public void run( ) {
        lm.removeUpdates(locationListenerGps);
        lm.removeUpdates(locationListenerNetwork);

        Location net_loc = null, gps_loc = null;
        if (gps_enabled)
            gps_loc = lm.getLastKnownLocation(LocationManager.GPS_PROVIDER);
        if (network_enabled)
            net_loc = lm
.getLastKnownLocation(LocationManager.NETWORK_PROVIDER);

        if (gps_loc != null && net_loc != null) {
            if (gps_loc.getTime( ) > net_loc.getTime( ))
                locationResult.gotLocation(gps_loc);
            else
                locationResult.gotLocation(net_loc);
            return;
        }

        if (gps_loc != null) {
```

```
                    locationResult.gotLocation(gps_loc);
                    return;
            }
            if (net_loc != null) {
                    locationResult.gotLocation(net_loc);
                    return;
            }
            locationResult.gotLocation(null);
        }
    }
    public static abstract class LocationResult {
            public abstract void gotLocation(Location location);
    }
}
```

回到自行开发的项目程序中处理。

先实作出LocationResult。

```
LocationResult locationResult = new LocationResult( ) {
    @Override
    public void gotLocation(Location location) {
            nowLat = location.getLatitude( );
            nowLng = location.getLongitude( );
    }
}
```

说明：
- nowLat 为事先定义的double变量，表示纬度。
- nowLng 为事先定义的double变量，表示经度。

先建构出MyLocation对象实体，并调用getLocation( )方法。

```
MyLocation myLocation = new MyLocation( );
myLocation.getLocation(this, locationResult);
```

就可以开始进行Location对象实体数据的取得。

另外，也提供了一个位置对象实体的最佳比较方法，以及判断是否相同提供者的判断。

先定义时间差值（目前设定为2min）：

```
private static final int TWO_MINUTES = 1000 * 60 * 2;
```

① 方法 isBetterLocation( ) :

```
protected boolean isBetterLocation(Location location,
        Location currentBestLocation) {
    if (currentBestLocation == null) {
        // 默认以新抓到的位置为最佳位置对象
        return true;
    }
    // 检查修正后新位置是否较新
    long timeDelta = location.getTime( ) - currentBestLocation.getTime( );
    boolean isSignificantlyNewer = timeDelta > TWO_MINUTES;
    boolean isSignificantlyOlder = timeDelta < -TWO_MINUTES;
    boolean isNewer = timeDelta > 0;
    // 判断是否距离上次位置时间为2min以上，是的话就对了，因为用户应该是在移动
状态
    if (isSignificantlyNewer) {
        return true;
    } else if (isSignificantlyOlder) {
        return false;
    }
    int accuracyDelta = (int) (location.getAccuracy( ) - currentBestLocation
            .getAccuracy( ));
    boolean isLessAccurate = accuracyDelta > 0;
    boolean isMoreAccurate = accuracyDelta < 0;
    boolean isSignificantlyLessAccurate = accuracyDelta > 200;
    // 检查两个位置的取得是否相同的提供者
    boolean isFromSameProvider = isSameProvider(location.getProvider( ),
            currentBestLocation.getProvider( ));
    if (isMoreAccurate) {
        return true;
    } else if (isNewer && !isLessAccurate) {
        return true;
    } else if (isNewer && !isSignificantlyLessAccurate
```

```
                && isFromSameProvider) {
            return true;
        }
        return false;
    }
```

② 方法isSameProvider( )：

```
private boolean isSameProvider(String provider1, String provider2) {
    if (provider1 == null) {
        return provider2 == null;
    }
    return provider1.equals(provider2);
}
```

## 11-2 基本Google Map

　　Google 自2012年12月3日起不再支持Maps API v1；而到了2013年3月3日前仍接受申请Maps API v1 key。Maps API v2改用com.google.android.gms.maps.MapFragment，仍需申请Maps API v2 key；Google Maps API v3 则改为Javascript API方式，使用WebView来显示Google Map，应用程序不需再申请Google Maps API key，但是对于地图呈现的API仍要使用API Key，其使用限制为同一页面免费存取次数为25000次/天。

因此本书将重点直接放在v3的版本的开发模式。

## 11-2-1　开发前期作业

① 注册申请Google账号，以该账号继续申请开发者身份。

② 前往https://code.google.com/apis/console注册之后，进入以下网页，并点选左侧的Services。

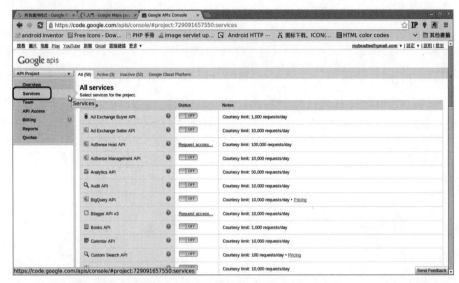

③ 右侧网页往下移动，直到看到Google Maps API v3。

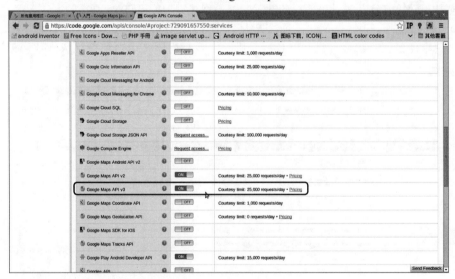

④ 将其开关切到ON即可，接着点选左侧的API Access项目，将看到如下的数据画面。

在 Simple API Access 中的 API Key 的数据就是开发上要用的 API Key。

## 11-2-2 Hello，Map

先在网页文件中进行测试，利用官方开发网站的"Hello，Map"来实测。

showmap.html

```
<!DOCTYPE html>
<html>
<head>
<meta name="viewport" content="initial-scale=1.0，user-scalable=no" />
<style type="text/css">
    html {
        height: 100%
    }
    body {
        height: 100%;
        margin: 0;
        padding: 0
    }
    #map_canvas {
        height: 100%
    }
</style>
```

```
<script type="text/javascript"
    src="http://maps.googleapis.com/maps/api/js?v=3&key=API Key放在此处
&sensor=false">
</script>
<script type="text/javascript">
    var map;
    function initialize( ) {
        var mapOptions = {
            center: new google.maps.LatLng(-34.397, 150.644),
            zoom: 8,
            mapTypeId: google.maps.MapTypeId.ROADMAP
        };
        map = new google.maps.Map(document.getElementById("map_canvas"), mapOptions);
    }
</script>
</head>
<body onload="initialize( )">
    <div id="map_canvas" style="width: 100%; height: 100%"></div>
</body>
</html>
```

笔者针对map的对象变量稍作修改，将其定义为全局变量，因为后续还会有其他JavaScript的自定义函数要对该变量进行相关存取设定。

通过浏览器开启，应该会看到如下澳大利亚悉尼的地图。

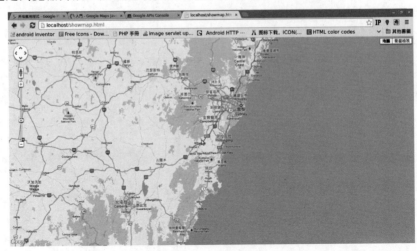

至此就算是完成了基本的前期作业。

## 11-2-3 在Android上开发的应用

因为要取得相关的地图数据是从因特网通信传输的,所以先在开发项目上设定权限INTERNET。而在版面配置上以WebView组件进行处理。

可将含有Google Map的html文件,放在远程服务器,或是项目资源中。

先来实作远程服务器的处理模式,假设放在http://android.ez2test.com/showmap.html。

```
private void initWebView( ) {
    webview.setWebViewClient(new WebViewClient( ));
    WebSettings settings = webview.getSettings( );
    settings.setJavaScriptEnabled(true);
    webview.addJavascriptInterface(new FromJavaScript(this), "Android");
    webview.loadUrl("http://android.ez2test.com/showmap.html");
    webview.setVisibility(View.INVISIBLE);
}
```

而如果放在项目资源中的处理方式,可以将html文件放在项目目录下的assets/子目录中。

```
private void initWebView( ) {
    webview.setWebViewClient(new WebViewClient( ));
    WebSettings settings = webview.getSettings( );
    settings.setJavaScriptEnabled(true);
    webview.addJavascriptInterface(new FromJavaScript(this), "Android");
    webview.loadUrl("file:///android_asset/showmap.html");
    webview.setVisibility(View.INVISIBLE);
}
```

在移动装置上呈现如下的画面(范例文件修改为取得定位信息):

接下来通常开发的重点分成三个部分：

• 以JavaScript来处理用户与地图对象的互动，例如点按特定位置出现标记（Marker）。

• JavaScript将用户的地图相关数据，传递给Android继续处理。例如用户的标记传递给Android的SQLite数据库储存。

• Android将信息数据传递给JavaScript，例如上一小节取得的定位经纬度数据传给JavaScript，并使地图置中于定位位置。

## 11-3 进阶Google Map

接着针对上一小节的重点开发进行实作。

### 11-3-1 JavaScript处理说明

声明及使用Google Maps API v3：

```
<script type="text/javascript"
    src="http://maps.googleapis.com/maps/api/js?v=3&key=API Key&sensor=false">
</script>
```

定义变量 map：

```
var map;
```

而 google.maps.Map 对象的建构会使用到 google.maps.MapOptions 对象，因此先建构：

```
var mapOptions = {
 center: new google.maps.LatLng(-34.397, 150.644),
 zoom: 14,
 mapTypeId: google.maps.MapTypeId.ROADMAP};
```

设定三个基本属性：
- center：定位中心位置。
- zoom：缩放值。
- mapTypedId：地图类型。

定位中心位置为 google.maps.LatLng 的经纬度类别对象实体，所以直接传入纬度与经度的数值来建构出来。缩放值越大越接近地面；越小越远离地面。

以下是 Google Maps API 提供的地图类型：
- MapTypeId.ROADMAP：会显示默认道路地图检视画面。
- MapTypeId.SATELLITE：会显示 Google 地球卫星影像。
- MapTypeId.HYBRID：会显示一般检视和卫星检视的混合画面。
- MapTypeId.TERRAIN：会根据地形信息显示实际地图。

建立最重要的 map 对象实体：

```
map = new google.maps.Map(document.getElementById("map_canvas"), mapOptions);
```

试着练习如何处理地图事件，当用户按下地图特定位置，使其增加一个标记（Marker）在地图上面。

先开发一段 JavaScript 的自定义函数 addMarker( )：

```
function addMarker(latLng){
    new google.maps.Marker({
        map: map,
        position: latLng
    });
}
```

回到 initialize( ) 中增加上事件处理，针对 map 发生 click 事件来调用 addMarker( )：

```
function initialize( ) {
    var mapOptions = {
      center: new google.maps.LatLng(-34.397, 150.644),
      zoom: 14,
      mapTypeId: google.maps.MapTypeId.ROADMAP
    };
    map = new google.maps.Map(document.getElementById("map_canvas"), mapOptions);
    google.maps.event.addListener(map, "click", function(event){
        addMarker(event.latLng);
    });
}
```

如此一来用户就可以触摸特定位置并标记。

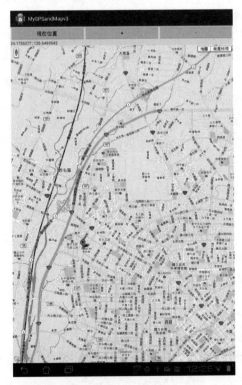

### ■ 11-3-2　JavaScript数据传回Android

先在 Android 程序中开发自定义类别：

```
private class FromJavaScript {
    private Context c;
    FromJavaScript(Context c) {
    this.c = c;
    }
    public void getLatLng(String data) {
        Log.i("brad", data);
    }
}
```

定义一个方法 getLatLng( )，传递的字符串参数就是从 JavaScript 传来的值。方法中将该字符串数据以 Log( ) 显示。

使 webview 增加该对象的处理界面，并定义 Android 字符串为调用的对象名称：

```
webview.addJavascriptInterface(new FromJavaScript(this), "Android");
```

回到 JavaScript 中的 addMarker( )：

```
function addMarker(latLng){
    new google.maps.Marker({
        map: map,
        position: latLng
    });
    Android.getLatLng(latLng.lat( ) + " : " + latLng.lng( ));
}
```

以对象变量名称为"Android"对象来调用 getLatLng( ) 方法。这样就可以将用户点击的特定位置，先在 map 上面标记，并传回给 Android 继续处理。

## ■ 11-3-3 以 Android 传递数据给 JavaScript

这种情况很多时候发生在对 JavaScript 的自定义函数进行调用，所以先定义一些自定义函数出来，移到指定中心位置：

```
function moveTo(lat, lng){
    var newpos = new google.maps.LatLng(lat, lng);
    map.panTo(newpos);
}
```

放大视图：

```javascript
function zoomIn( ){
    var zoom = map.getZoom( );
    if (zoom<21) zoom++;
    map.setZoom(zoom);
}
```

缩小视图：

```javascript
function zoomOut( ){
    var zoom = map.getZoom( );
    if (zoom>1) zoom--;
    map.setZoom(zoom);
}
```

回到Android中，只需要调用webview的loadUrl( )即可，改回之前以GPS取得定位：

```java
LocationResult locationResult = new LocationResult( ) {
    @Override
    public void gotLocation(Location location) {
        nowLat = location.getLatitude( );
        nowLng = location.getLongitude( );
        info.setText(nowLat + " : " + nowLng);
        webview.loadUrl("javascript:moveTo(" + nowLat + "," + nowLng + ")");
    }
}
```

或是

```java
webview.loadUrl("javascript:zoomIn( )");
webview.loadUrl("javascript:zoomOut( )");
```

# Chapter 12

## 第12课　传感器运行原理及应用

12-1　传感器运行原理与应用

12-2　三轴加速传感器

12-3　重力加速度传感器

12-4　磁极方向传感器

12-5　光线/温度/湿度/压力传感器

## 12-1 传感器运行原理与应用

### ■ 12-1-1 基本概念

事实上这个章节所要讨论的传感器 Sensor，指的是通过 android.hardware.Sensor 的 API 与实际硬件的传感器两者之间所响应的数据。

在目前的移动装置上面，或多或少都配备一些传感器设备。而硬件配备与否，先天就决定了这个移动装置是否可以支持使用该项传感器装置；后天上还需要软件的支持，也就是其 Android API Level 是否支持，两相搭配之下，才能通过该项装置的传感器来取得相关的数据。

原则上可以分成三种类型的传感器：

① 位移传感器
- 三轴加速传感器。
- 重力加速度传感器。
- 陀螺仪传感器。
- 线性加速度传感器。
- 旋转向量传感器。

② 环境传感器
- 光线传感器。
- 温度传感器。
- 相对湿度传感器。
- 大气压力传感器。

③ 位置传感器
- 方向传感器。
- 磁极传感器。
- 接近传感器。

### ■ 12-1-2 处理原则

处理的原则如下：

- 调用 getSystemService( ) 方法，传递 SENSOR_SERVICE，取得 android.hardware.SensorManager 的对象实体。
- 再由 SensorManager 对象实体调用 getDefaultSensor( ) 方法，传递指定的 Sensor 型态的 int 值，传回 Sensor 对象实体。
- 如果传回的 Sensor 对象实体为 null，就表示不支持该项传感器。
- 建立一个自定义类别实作 android.hardware.SensorEventListener 界面。

- Override 其 onSensorChange( )方法。
- 将传递来的 SensorEvent 对象实体，取得该传感器的数据。
- 最后由 SensorManager 对象实体调用 registerListener( )方法，设定 SensorEventListener 对象及周期频率即可。
- 当不再使用该传感器时，要记得 SensorManager 对象实体调用 unregisterListener( )方法。

## ■ 12-1-3 实作开发

声明对象变量：

```
static SensorManager smgr;
private Sensor sensor;
private boolean isSensorEnable;
```

侦测是否支持：

```
sensor = smgr.getDefaultSensor(Sensor.TYPE_ACCELEROMETER);
// 侦测指定的传感器是否支持
if (sensor != null) {
    isSensorEnable = true;
}
```

开发自定义类别处理传感器事件监听器：

```
private class MySensorEventListener implements SensorEventListener {
    // 精准度改变事件
    @Override
    public void onAccuracyChanged(Sensor sensor, int accuracy) {
    }
    // 感应事件
    @Override
    public void onSensorChanged(SensorEvent event) {
    }
}
```

注册与解除注册：

```
@Override
protected void onResume( ) {
```

```
    super.onResume( );
    if (isSensorEnable) {
        smgr.registerListener(listener, sensor,
                    SensorManager.SENSOR_DELAY_UI);
    }
}
@Override
protected void onPause( ) {
    super.onPause( );
    if (listener != null) {
        smgr.unregisterListener(listener);
    }
}
```

传感器监听的频率设定如下：
- SensorManager.SENSOR_DELAY_NORMAL: 0.2s。
- SensorManager.SENSOR_DELAY_UI: 0.06s。
- SensorManager.SENSOR_DELAY_GAME: 0.02s。
- SensorManager.SENSOR_DELAY_FASTEST: 0s。
- API Level 11+可以自行指定微秒单位。

开发的重点在于自定义类别处理传感器事件监听器中，以 SensorEvent 对象实体调用其属性 values，传回一个 float[] 数组的值。

### ■ 12-1-4 用户装置支持处理

对于开发者而言，既然存在这样的状况，而又不想增加开发上的考虑，则可以在 Androidmanifest.xml 文件中针对特定的传感器，设定其属性值。使得用户如果通过 Google play 的市场下载安装之前，可以得知用户的移动装置是否支持 App 中的需求。通常这样的处理方式是 App 中对于该项传感器装置是必要条件，也就是说如果没有该项传感器装置，就完全无法使用 App 的状况。如果没有该项特定的传感器装置仍有其他方式可以替代，则不建议如此进行设定。

假设要侦测的传感器为三轴加速传感器，则如下设定在 AndroidManifest.xml：

```
<uses-feature
    android:name="android.hardware.sensor.accelerometer"
    android:required="true" />
```

其他：

- 三轴加速传感器：accelerometer。
- 环境温度/湿度传感器：barometer。
- 方向罗盘传感器：compass。
- 陀螺仪传感器：gyroscope。
- 光线传感器：light。
- 接近传感器：proximity。

## 12-2　三轴加速传感器

三轴加速传感器是利用移动装置所在的空间坐标而感应其移动加速度的传感器。

当移动装置对于用户以直向握持的状态而言，左右移动则为 $X$ 坐标轴，前后移动则为 $Y$ 坐标轴，上下移动则为 $Z$ 坐标轴。其值的单位为每秒平方所移动的公尺距离。

而其各坐标轴的值，是通过 SensorEvent.values 属性取得的 float[] 数组来的：

- SensorEvent.values[0] => $X$ 坐标轴。
- SensorEvent.values[1] => $Y$ 坐标轴。
- SensorEvent.values[2] => $Z$ 坐标轴。

接下来直接进行开发实作。实作方式不打算将数据直接呈现出来，因为即使是敏感度较差的状况，也将会是每 0.2s 更新数据，不易判断处理；甚至将数据以数学方式重新整理，也只是数字。因此笔者以自定义 View 来绘制几何图形呈现，应该会比较合适。

通常三轴加速传感器的 $XY$ 坐标轴较为常被应用，因此以下实作以 $XY$ 坐标为主。

先进行基本侦测，传递参数为 Sensor.TYPE_ACCELEROMETER：

```
sensor = smgr.getDefaultSensor(Sensor.TYPE_ACCELEROMETER);
// 侦测指定的传感器是否支持
if (sensor != null) {
    isSensorEnable = true;
    listener = new MySensorEventListener( );
}
```

建立一个自定义 View，想要将三轴加速传感器反映在一个几何圆上面。

```
private class TestView extends View {
    private int viewW, viewH;
    private boolean isInited;
    private Paint paintText, paintValue;
```

```java
        float cx, cy, cr;
    public TestView(Context context) {
            super(context);
            setBackgroundColor(Color.BLACK);

            paintText = new Paint( );
            paintText.setTextSize(36);
            paintText.setColor(Color.YELLOW);

            paintValue = new Paint( );
            paintValue.setTextSize(36);
            paintValue.setColor(Color.RED);
    }
    private void init( ) {
            viewW = getWidth( );
            viewH = getHeight( );

            cx = viewW / 2;
            cy = viewH / 2;
            cr = 50;

            isInited = true;
    }
    @Override
    protected void onDraw(Canvas canvas) {
            if (!isInited) {
                    init( );
                    init( )
                    canvas.drawText(
                            isSensorEnable ? "三轴加速器支持" + rotation : "装
                                            置不支持",
                            viewW / 4, viewH / 4, paintText);
            }
            canvas.drawCircle(cx, cy, cr, paintText);
    }
}
```

重点说明：
- 背景设定为全黑。

- 画一个几何圆，一开始的位置在正中央，半径为50像素。

所以在onCreate( )方法中整理如下：

```
@Override
protected void onCreate(Bundle savedInstanceState) {
    super.onCreate(savedInstanceState);
    tv = new TestView(this);
    setContentView(tv);
    rotation = getWindowManager( ).getDefaultDisplay( ).getRotation( );
    smgr = MainActivity.smgr;
    sensor = smgr.getDefaultSensor(Sensor.TYPE_ACCELEROMETER);
    // 侦测指定的传感器是否支持
    if (sensor != null) {
        isSensorEnable = true;
        listener = new MySensorEventListener( );
    }
}
```

开发自定义传感器事件监听器：

```
private class MySensorEventListener implements SensorEventListener {
    // 精准度改变事件
    @Override
    public void onAccuracyChanged(Sensor sensor, int accuracy) {
    }
    // 感应事件
    @Override
    public void onSensorChanged(SensorEvent event) {
        x = event.values[0] * 7;
        y = event.values[1] * 7;
        z = event.values[2];
        if (rotation == 0) {
            sx = x;
            sy = y;
            sz = z;
        } else if (rotation == 3) {
```

```
                sx = y;
                sy = -1 * x;
                sz = z;
            }
            tv.addCX(-1 * sx);
            tv.addCY(sy);
            tv.postInvalidate( );
        }
    }
```

重点说明：

tv 为 TestView 的对象实体，若有改变数据，则调用 TestView 重新绘制处理传感器事件监听器的注册与解除：

```
@Override
protected void onResume( ) {
    super.onResume( );
    if (isSensorEnable) {
        smgr.registerListener(listener, sensor,
                    SensorManager.SENSOR_DELAY_UI);
    }
}
@Override
protected void onPause( ) {
    super.onPause( );
    if (listener != null) {
        smgr.unregisterListener(listener);
    }
}
```

其中特别的地方是有一个 rotation 的 int 变量，其值是通过调用 getWindowManager( ).getDefaultDisplay( ).getRotation( ) 而来。在本单元的范例是以垂直方向执行，而在一般手机的状态下，其值为 0，但是相同的设定下，在平板电脑（以 TF-101 为例）下因为预设为水平方向，所以得到其值为 3，所以要将 XY 坐标轴作调整转换，这样可以使该应用程序适用于不同的移动装置。

正常执行下应该会看到一个几何圆形，会随着用户前后左右翻转而移动，可以应用在控制方向的应用。

## 12-3 重力加速度传感器

重力加速度传感器是利用移动装置所在的空间坐标而感应其重力加速度的传感器（Sensor.TYPE_GRAVITY）。

坐标轴的观念与三轴加速度传感器是相同的，而其静置状态下，Z坐标轴的值大约在9.8，也就是一个g值。

一样是以一个几何图形的圆来呈现其效果，当感应侦测Z坐标轴的改变时，会实时变更圆的半径大小。当用户晃动移动装置的同时，就会变更圆的大小，若是整个移动装置反面过来，则会得到负值，使得原本是黄色的圆，转变成为红色的圆。

应用举例：
- 数值变化为晃动手机，模拟为钓鱼甩竿。
- 数值负数为翻面，触发的同时暂停游戏进行，或是来电铃声停止。

```java
package tw.brad.book.mysensor;

import android.app.Activity;
import android.content.Context;
import android.graphics.Canvas;
import android.graphics.Color;
import android.graphics.Paint;
import android.hardware.Sensor;
import android.hardware.SensorEvent;
import android.hardware.SensorEventListener;
import android.hardware.SensorManager;
import android.os.Bundle;
import android.view.View;

public class GravityTest extends Activity {
    private TestView tv;
    private SensorManager smgr;
    private Sensor sensor;
    private boolean isSensorEnable;
    private int rotation;
    private MySensorEventListener listener = null;
    private float x, y, z, sx, sy, sz;

    @Override
```

```java
    protected void onCreate(Bundle savedInstanceState) {
        super.onCreate(savedInstanceState);
        tv = new TestView(this);
        setContentView(tv);
        rotation = getWindowManager( ).getDefaultDisplay( ).getRotation( );
        smgr = MainActivity.smgr;
        sensor = smgr.getDefaultSensor(Sensor.TYPE_GRAVITY);
        // 侦测指定的传感器是否支持
        if (sensor != null) {
            isSensorEnable = true;
            listener = new MySensorEventListener( );
        }
    }

    @Override
    protected void onResume( ) {
        super.onResume( );
        if (isSensorEnable) {
            smgr.registerListener(listener, sensor,
                SensorManager.SENSOR_DELAY_UI);
        }
    }

    @Override
    protected void onPause( ) {
        super.onPause( );
        if (listener != null) {
            smgr.unregisterListener(listener);
        }
    }

    private class MySensorEventListener implements SensorEventListener {
        // 精准度改变事件
        @Override
        public void onAccuracyChanged(Sensor sensor, int accuracy) {
        }

        // 感应事件
        @Override
```

```java
        public void onSensorChanged(SensorEvent event) {
            x = event.values[0] * 100;
            y = event.values[1] * 100;
            z = event.values[2];
            if (rotation == 0){
                sx = x;
                sy = y;
                sz = z;
            }else if (rotation == 3)
                {
                sx = -1*y;
                sy = -1*x;
                sz = z;
            }
            tv.cr = z * 10 + -30;
            tv.postInvalidate( );
        }
    }
    private class TestView extends View
        private int viewW, viewH;
        private boolean isInited;
        private Paint paintText, paintValue;
        float cx, cy, cr;
        public TestView(Context context) {
        super(context);
        setBackgroundColor(Color.BLACK);
        paintText = new Paint( );
        paintText.setTextSize(36);
        paintText.setColor(Color.YELLOW);
        paintValue = new Paint( );
        paintValue.setTextSize(36);
        paintValue.setColor(Color.RED);
        }
    private void init( ) {
        viewW = getWidth( );
```

```
            viewH = getHeight( );
            cx = viewW / 2;
            cy = viewH / 2;
            cr = 50;
            isInited = true;
    }
    @Override
    protected void onDraw(Canvas canvas) {
        if (!isInited) {
            init( );
            canvas.drawText(isSensorEnable?" 重力加速度传感器支持" :
                                            "装置不支持",
                viewW / 8, viewH / 8, paintText);
        }
        canvas.drawText((int) sx + ":" + (int) sy + ":" + sz,
            viewW / 8,viewH / 8, paintValue);
        canvas.drawCircle(cx, cy, Math.abs(cr), cr > 0 ?
            paintText: paintValue);
        }
     }
}
```

## 12-4 磁极方向传感器

磁极方向传感器是利用移动装置所在的空间坐标而感应其各自坐标的磁场强弱值的传感器（Sensor.TYPE_MAGNETIC_FIELD）。

在各坐标轴的数值就是移动装置目前侦测的磁场磁力的数据，磁场磁力最强的就是地球N极，也就是向北方向，所以若移动装置平放的状态下，表示Z坐标轴为固定，而以XY坐标轴所接收的数据可以呈现出一个基本的指北针。

以下范例中会将抓到的XY坐标值放大4倍，以便容易处理明显的数据，而数学几何Y坐标轴与屏幕的Y轴刚好相反（屏幕向上为负方向，向下为正方向），所以乘以−4。先在屏幕中间绘出数学几何的XY坐标轴及中心点，而黄色线条从中心点出发所绘制的就是目前磁极的北极。

```
package tw.brad.book.mysensor;

import android.app.Activity;
import android.content.Context;
import android.graphics.Canvas;
import android.graphics.Color;
import android.graphics.Paint;
import android.hardware.Sensor;
import android.hardware.SensorEvent;
import android.hardware.SensorEventListener;
import android.hardware.SensorManager;
import android.os.Bundle;
import android.view.View;

public class MagneticFieldTest extends Activity {
    private TestView tv;
    private SensorManager smgr;
    private Sensor sensor;
    private boolean isSensorEnable;
    private int rotation;
    private MySensorEventListener listener = null;
    private float x, y, z, sx, sy, sz;

    @Override
    protected void onCreate(Bundle savedInstanceState) {
        super.onCreate(savedInstanceState);
        tv = new TestView(this);
        setContentView(tv);
```

```java
        rotation = getWindowManager( ).getDefaultDisplay( ).getRotation( );
    smgr = MainActivity.smgr;
    sensor = smgr.getDefaultSensor(Sensor.TYPE_MAGNETIC_FIELD);
        // 侦测指定的传感器是否支持
        if (sensor != null) {
            isSensorEnable = true;
            listener = new MySensorEventListener( );
        }
}

@Override
protected void onResume( ) {
    super.onResume( );
    if (isSensorEnable) {
        smgr.registerListener(listener, sensor,
            SensorManager.SENSOR_DELAY_UI);
    }
}

@Override
protected void onPause( ) {
    super.onPause( );
    if (listener != null) {
        smgr.unregisterListener(listener);
    }
}

  private class MySensorEventListener implements SensorEventListener {
      private boolean isNotFirst;
      private float zFirst;

      // 精准度改变事件
      @Override
      public void onAccuracyChanged(Sensor sensor, int accuracy) {
      }
      // 感应事件
      @Override
      public void onSensorChanged(SensorEvent event) {
          x = event.values[0]*4;
          y = event.values[1]*-4;
```

```
                z = event.values[2];
                if (rotation == 0) {
                    sx = x;
                    sy = y;
                    sz = z;
                else if (rotation == 3)
                    sx = -1 * y;
                    sy = -1 * x;
                    sz = z;

                    tv.cr = 30;
                    tv.postInvalidate( );
                }
    }
    private class TestView extends View {
        private int viewW, viewH;
        private boolean isInited;
        private Paint paintText, paintValue;
        float cx, cy, cr;
        public TestView(Context context) {
            super(context);
            setBackgroundColor(Color.BLACK);

            paintText = new Paint( );
            paintText.setTextSize(36);
            paintText.setColor(Color.YELLOW);

            paintValue = new Paint( );
            paintValue.setTextSize(36);
            paintValue.setColor(Color.RED);
        }
            private void init( ) {
            viewW = getWidth( );
            viewH = getHeight( );
            cx = viewW / 2;
            cy = viewH / 2;
            cr = 30;
```

```
                isInited = true;
            }
            @Override
            protected void onDraw(Canvas canvas) {
                if (!isInited) {
                    init( );
                    canvas.drawText(isSensorEnable ?
                            "磁极方位传感器支持" : "装置不支持",
                        viewW / 8, viewH / 8, paintText);
                }
                canvas.drawText((int) sx + ":" + (int) sy + ":" + sz,
                    viewW / 8, viewH / 8, paintValue);
                canvas.drawCircle(cx , cy , 4, paintValue);
                canvas.drawLine(0, cy, viewW, cy, paintValue);
                canvas.drawLine(cx, 0, cx, viewH, paintValue);
                canvas.drawLine(cx, cy, cx +sx, cy + sy, paintText);
            }
        }
}
```

## 12-5 光线/温度/湿度/压力传感器

这部分都是用来测量环境的传感器：
- 光线传感器（Sensor.TYPE_LIGHT）。
- 温度传感器（Sensor.TYPE_AMBIENT_TEMPERATURE）。
- 相对湿度传感器（Sensor.TYPE_RELATIVE_HUMIDITY）。
- 大气压力传感器（Sensor.TYPE_RELATIVE_PRESSURE）。

只有一个SensorEvent.values[0]值传回，值越大光度越亮；值越小越暗。通常可以通过该数据调整App中的相关属性。其他的传感器只是所代表的值不同而已，其他逻辑架构都非常类似。

```
package tw.brad.book.mysensor;

import android.app.Activity;
import android.content.Context;
import android.graphics.Canvas;
```

```java
import android.graphics.Color;
import android.graphics.Paint;
import android.hardware.Sensor;
import android.hardware.SensorEvent;
import android.hardware.SensorEventListener;
import android.hardware.SensorManager;
import android.os.Bundle;
import android.view.View;
public class LightTest extends Activity {
    private TestView tv;
    private SensorManager smgr;
    private Sensor sensor;
    private boolean isSensorEnable;
    private MySensorEventListener listener = null;
    private float x, y, z, sx, sy, sz;

    @Override
    protected void onCreate(Bundle savedInstanceState) {
        super.onCreate(savedInstanceState);
        tv = new TestView(this);
        setContentView(tv);

        smgr = MainActivity.smgr;
        sensor = smgr.getDefaultSensor(Sensor.TYPE_LIGHT);
        // 侦测指定的传感器是否支持
        if (sensor != null) {
            isSensorEnable = true;
            listener = new MySensorEventListener( );
        }
    }

    @Override
    protected void onResume( ) {
        super.onResume( );
        if (isSensorEnable) {
            smgr.registerListener(listener, sensor,
                    SensorManager.SENSOR_DELAY_UI);
        }
    }
```

```java
        @Override
        protected void onPause( ) {
                super.onPause( );
                if (listener != null) {
                        smgr.unregisterListener(listener);
                }
        }

    private class MySensorEventListener implements SensorEventListener {
            // 精准度改变事件
            @Override
            public void onAccuracyChanged(Sensor sensor, int accuracy) {
            }

            // 感应事件
            @Override
            public void onSensorChanged(SensorEvent event) {
                    x = event.values[0];
                    y = event.values[1];
                    z = event.values[2];

                    sx = x;
                    sy = y;
                    sz = z;

                    tv.cr = z * 10 + 30;
                    tv.postInvalidate( );
            }
    }

    private class TestView extends View {
            private int viewW, viewH;
            private boolean isInited;
            private Paint paintText, paintValue;
            float cx, cy, cr;

            public TestView(Context context) {
                    super(context);
                    setBackgroundColor(Color.BLACK);

                    paintText = new Paint( );
                    paintText.setTextSize(36);
```

```
        paintText.setColor(Color.YELLOW);

        paintValue = new Paint( );
        paintValue.setTextSize(36);
        paintValue.setColor(Color.RED);
    }

    private void init( ) {
        viewW = getWidth( );
        viewH = getHeight( );

        cx = viewW / 2;
        cy = viewH / 2;
        cr = 50;

        isInited = true;
    }

    @Override
    protected void onDraw(Canvas canvas) {
        if (!isInited) {
            init( );
            canvas.drawText(isSensorEnable ?
                    "光线传感器支持" : "装置不支持",
                viewW / 8, viewH / 8, paintText);
        }

        canvas.drawText(sx + ":" + sy + ":" + sz, viewW / 8,
            viewH / 8, paintValue);
        canvas.drawCircle(cx, cy, Math.abs(cr), cr > 0 ?
            paintText: paintValue);
        }
    }
}
```

# MEMO

# 13
Chapter

## 第13课　资源与国际化

13-1　提供资源内容

13-2　存取资源内容

13-3　应用程序执行中的改变

13-4　资源内容的区域化

先来看一段Java程序最熟悉不过的"Hello，World"再说：

```
……
System.out.println("Hello, World");
……
```

对于这段Java程序而言，"Hello，World"就是一种资源，算是文字信息的资源。但是，无论在任何语系的操作系统下，通通看到的都是"Hello，World"。如果希望可以在中文环境下看到"您好，全世界"，甚至于因为登入账号的用户为brad，当执行这段程序时可以看到"您好，brad"。这样的"Hello，World"程序应该会更为广泛地应用。因此，最简单的方式就是将准备呈现的资源内容给独立在程序代码以外的地方，独立进行维护及管理。

通常在开发应用程序的时候，会将应用程序中所用的影像或是文字信息等相关资源独立维护，而不会直接写在应用程序代码之中，因为那么做只会增加日后维护的困扰，除非你觉得这个应用程序项目只需要写这么一次就可以为你赚很多钱。而将应用程序代码以外的资源独立的处理方式，除了提高日后的维护性之外，还可以进一步地发展成为多国语系的应用程序，使该项目的应用范畴更为广阔。Android项目的资源内容通常是放在项目目录之下的res/子目录，这个章节就是准备来深入了解资源子目录。

针对任何类型的资源内容，都会有预设及其他多重的替代性资源可以选择使用，通常有以下的原则来处理：

• 默认的资源内容应用在装置的组态设定或是没有其他替代性资源可以使用的时机。

• 替代性资源内容会是完整的一组内容，特别用在特定的场合,通常会放在目录名称容易辨认的子目录下。

举例来说，一般的用户界面的版面配置内容会放在res/layout/子目录下，通常默认是针对移动装置拿直的方式来操作，或是将移动装置横向（landscape）操作，如果在使用界面上想要另外处理，那可能就会另外建立出一个子目录res/layout-land/来存放当用户横向使用应用程序的资源内容。

上图的Android应用程序分别提供给装置A及装置B使用，因为只有处理预设资源，所以无论是装置A的直向操作或是装置B的横向操作，所看的用户界面内容

都是一样的，但是呈现的整体比例却是不同的，所产生的效果就不是很恰当。

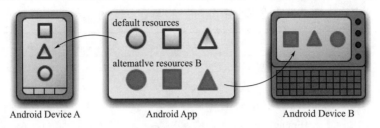

上图在Android应用程序开发架构中，除了预设资源内容的提供之外，还另外提供替代性资源内容给横向操作的装置使用。因此，无论是装置A或是装置B都可以看到比例上或是视觉上较为一致的用户界面。其实，除了视觉上之外的应用，还可以运用在信息内容的处理。

上述说明只是让读者了解资源使用的重要性。以下将会提供读者如何架构完善的资源内容，以及运用替代性资源内容的手法及时机。

## 13-1 提供资源内容

在Android项目架构下，预设建立的项目目录结构中，就已经将资源内容独立在res/的子目录下：

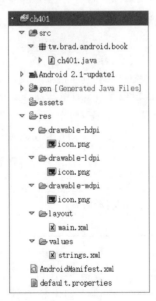

### ■ 13-1-1 预设资源内容及架构

目前预设的项目建立的同时，产生了res/drawable-xxxx/、res/layout/ 及 res/values/3

个子目录。xxxx表示分辨率不同所使用不同的绘图资源目录，hdpi为高分辨率，ldpi为低分辨率，mdpi则是适用于中分辨率的装置。较为完整的内容如下表所示。

| 子目录 | 存放资源内容格式及型态 |
| --- | --- |
| res/anim/ | • 动画资源内容<br>• xml文件格式<br>• Tween Animation<br>• Frame Animation |
| res/color/ | • 定义颜色值<br>• xml文件格式 |
| res/drawable-xxxx/ | • 影像图资源<br>• 点阵图文件格式或是xml文件格式<br>• .png<br>• .9.png<br>• .jpg<br>• .gif<br>• .xml |
| res/layout/ | • 用户界面的版面配置资源<br>• xml文件格式 |
| res/menu/ | • 菜单资源，应用在选项菜单(Options Menu)，本文菜单(Context Menu)或是子菜单(Sub Menu)<br>• xml文件格式 |
| res/raw/ | • 文本文件资源内容<br>• 任何没有经过压缩处理的原始的文本文件格式<br>• 如果在应用程序中只是单纯地读取文字数据，而并不需要以资源型态管理，则可以放在assets/子目录之下。其实这就是资源和资产的不同之处，简单的区分方式就是，资产总是固定的，资源是可以选择改变的内容 |
| res/values/ | • 可能是文字内容，也可能是数值数据等的值<br>• xml文件格式 |
| res/xml/ | • 任何形式的xml资源内容<br>• xml文件格式 |

通常不会直接将文件放进特定的子目录之下，而会通过特定的操作程序存放。这样才能够达到进行资源管理的目的。影像图文件就可以直接放进res/drawable-xxxx/的子目录下，放完之后，记得按下"F5"功能键。

## ■ 13-1-2　替代选择性资源内容

一个完善的Android项目应用程序都应该提供除了预设资源以外的其他替代选择性的资源内容。至少从用户的语系角度来看，就可以使更多的用户使用精心所设计出来的应用程序。可能有人会举这么一个例子来反驳，就是写一个卜卦程序，应该只有中文语系的用户会去使用，你仍然可以针对直向或是横向操作移动装置的不同习性来处理不同的版面配置，使得虽然只有中文语系的用户会使用的应用程序，仍然顾及到不同的操作习性的层面，这就是替代选择性预设资源的重要性。

处理的程序相当简单。先建立替代选择性的资源目录。

① 也是在项目的res/子目录之下，依照命名原则："资源项目-限定名称"，建立出类似像res/layout-land/或是res/drawable-hdpi/的子目录。而限定名称还必须有更细的区分，则可以继续描述命名下去，中间以 "-" (dash)做区隔即可。

② 替代选择性资源目录下的文件名应该与默认资源目录下的一致。

例如：

```
res/
    drawable/
        icon.png
        welcome.png
        background.png
    drawable-hdpi/
        icon.png
        welcome.png
        background.png
    drawable-ldpi/
        icon.png
        welcome.png
        background.png
    drawable-mdpi/
        icon.png
        welcome.png
        background.png
```

在以上架构中，drawable-xxxx/子目录下的文件内容及名称都与drawable/子目录下的一样，虽然所呈现的影像文件内容可能完全不一样。

以下列出Android目前常用的限定名称：

| 区分项目 | 限定名称范例 | 说明 |
|---|---|---|
| MCC及MNC | mcc466<br>mcc466-mnc01<br>mcc466-mnc92 | 移动装置国码（地区码）及网络码<br>这是定义在ITU E.212（Land Mobile Numbering Plan），用来识别无线电信网络的移动站。特别是GSM及UMTS网络。相关详细对照表在附录中<br>mcc466：表示台湾地区<br>mcc460-mnc01：表示台湾地区远传电信<br>mcc310-mnc92：表示台湾地区的中华电信 |
| 语言及区域 | en<br>en-rUS<br>en-rCA<br>zh-rTW | 限定在语系-区域的资源目录，使用两码的ISO 639-1的语系代码，及两码的ISO 3166-1-alpha-2的区域代码，相关详细对照表在附录中。大小写没有区分，如果有指定区域代码，必须前置"r"表示区域（region）<br>• en表示英文<br>• en-rUS表示英文-美国<br>• en-rCA表示英文-加拿大<br>• zh-rTW表示中文-台湾地区 |
| 屏幕尺寸 | small<br>normal<br>large | small应用在低分辨率的QVGA的屏幕上QVGA（240×320），2.6"-3.0"对角线normal应用在传统的中分辨率HVGA的屏幕上<br>• WQVGA (240×400)，3.2"-3.5"对角线<br>• FWQVGA (240×432)，3.5"-3.8"对角线<br>• HVGA (320×480)，3.0"-3.5"对角线<br>• WVGA (480×800)，3.3"-4.0"对角线<br>• FWVGA (480×854)，3.5"-4.0"对角线large应用在中分辨率VGA的屏幕上<br>• WVGA (480×800)，4.8"-5.5"对角线<br>• FWVGA (480×854)，5.0"-5.8"对角线 |
| 屏幕形状 | long<br>notlong | long(宽屏幕)：WQVGA，WVGA，FWVGA<br>notlong：QVGA，HVGA，VGA<br>这个项目并非是直向或是横向的操作屏幕，而是屏幕本体 |
| 屏幕操作方向 | port<br>land | Port垂直操作方向(直向)<br>land水平操作方向(横向) |
| 基座模式 | car<br>desk | car车用模式<br>desk桌上模式<br>适用于Android API Level 8以上 |
| 夜间模式 | night<br>notnight | night夜间模式<br>notnight白天模式<br>适用于Android API Level 8以上 |

续表

| 区分项目 | 限定名称范例 | 说明 |
|---|---|---|
| 屏幕分辨率 | ldpi<br>mdpi<br>hdpi<br>nodpi | ldpi低分辨率，大约在120dpi<br>mdpi中分辨率，大约在160dpi<br>hdpi高分辨率，大约在240dpi<br>nodpi应用在点阵资源内容，不限制在特定的解析度 |
| 触控屏幕模式 | notouch<br>stylus<br>finger | notouch没有触控屏幕装置<br>stylus使用触控笔模式<br>finger手指的触控屏幕 |

注意事项：

- 单一资源目录的限定名称可以多重指定，例如layout-zh-rTW-car。
- 使用多重限定名称的规则必须按照上表的顺序性使用。例如要建立一个高分辨率的垂直影像资源内容，则其名称为drawable-port-hdpi，若写成drawable-hdpi-port就是错误的方式。所以上表的顺序性是非常重要的参考依据。
- 各个多重限定名称资源目录不可以使用巢状结构。
- 也不允许使用特定资源但是多重限定名称，例如drawable-mdpi-hdpi想要同时支持中高分辨率的影像资源内容。

## 13-2 存取资源内容

一旦完成资源内容的配置及存放，接着下来就是要讨论如何存取使用这些资源内容。

只要释放在上一节中的资源内容，也就是存放在项目架构下的res/资目录资源内容，都可以通过其资源唯一识别码来进行存取。所有的资源内容都会被定义在R.java之中，读者不需要手动处理该文件内容，甚至于说是千万不要自行改变R.java中的内容，开发工具程序-aapt会自动配置这些资源内容并赋予一个该项目应用程序的唯一资源识别码。

一个资源识别码的组成内容会区分成两部分：资源型态（resource type）及资源名称（resource name）。

- 资源型态：例如string,drawable或是layout。
- 资源名称：例如影像文件名称或是自定义View组件的名称。

有两种方式来进行资源内容的存取。

- 程序代码：R.资源型态.资源名称，例如："R.string.message"。
- XML定义档：@资源型态/资源名称，例如："@string/message"。

## 13-2-1 程序代码中存取资源内容

在程序代码中使用指定资源内容时,只需要指定在 R 类别中的唯一识别码,通常使用其静态的成员属性名称的方式。假设有一个显示文字信息的组件 TextView,想要呈现文字信息的资源内容放在 res/values/strings.xml 中的 message。因此,就可以下段程序代码中的 R.string.message 来进行存取:

① res/values/strings.xml

```xml
<?xml version="1.0" encoding="utf-8"?>
<resources>
    <string name="app_name">ch401</string>
    <string name="hello">Hello World, ch401!</string>
    <string name="message">Hello, Brad</string>
</resources>
```

② res/layout/main.xml

```xml
<?xml version="1.0" encoding="utf-8"?>
<LinearLayout xmlns:android="http://schemas.android.com/apk/res/android"
    android:orientation="vertical"
    android:layout_width="fill_parent"
    android:layout_height="fill_parent"
    >
<TextView
     android:id="@+id/tv"
    android:layout_width="fill_parent"
    android:layout_height="wrap_content"
    android:text="@string/hello"
    />
</LinearLayout>
```

③ src/tw.brad.android.book/ch401.java

```java
package tw.brad.android.book;

import android.app.Activity;
import android.os.Bundle;
import android.widget.TextView;

public class ch401 extends Activity {
    private TextView tv = null;
```

```
    @Override
    public void onCreate(Bundle saved InstanceState) {
        super.onCreate(saved InstanceState);
        setContentView(R.layout.main);
        tv = (TextView)findViewById(R.id.tv);
        tv.setText(R.string.message);
    }
}
```

## ■ 13-2-2  XML中存取资源内容

在xml文件中存取资源内容的方法与在程序中不太一样，相当于说明该资源内容放在哪里，也就是会有1个前置符"@"，表示"在"（at）哪里的意思。其格式为"@资源型态/资源名称"，例如在main.xml文件中定义一个按钮组件上的文字android:text="@string/bt_click"，就直接延续上个范例来看：

① res/values/strings.xml

```
<?xml version="1.0" encoding="utf-8"?>
<resources>
    <string name="app_name">ch401</string>
    <string name="hello">Hello World, ch401!</string>
    <string name="message">Hello, Brad</string>
    <string name="bt_click">按下去吧</string>
</resources>
```

② res/layout/main.xml

```
<?xml version="1.0" encoding="utf-8"?>
<LinearLayout xmlns:android="http://schemas.android.com/apk/res/android"
    android:orientation="vertical"
    android:layout_width="fill_parent"
    android:layout_height="fill_parent"
    >
<TextView
     android:id="@+id/tv"
    android:layout_width="fill_parent"
    android:layout_height="wrap_content"
```

```xml
        android:text="@string/hello"
    />
<Button
    android:id="@+id/bt"
    android:layout_width="fill_parent"
    android:layout_height="wrap_content"
    android:text="@string/bt_click"
    />
</LinearLayout>
```

## 13-3 应用程序执行中的改变

当应用程序已经激活执行，可能因为用户的因素而改变装置的模式，最常见的状况就是原本垂直操作模式改成水平横向操作，或是从桌上基座拿到车上基座使用等执行中的状态模式改变。无论如何都会被重新启动（restart），也就是调用 onDestroy( ) 之后，再度重新启动 onCreate( )。这样的执行过程是自动被调用，以便应用新的装置模式。

常见的处理方式有两种，可以任选一种恰当的方式来实作。

### ■ 设计一个保留及回存对象

当重新激活 Activity 的生命周期的时候，原本使用的对象状态属性等数据，必须在还没有被"摧毁"之前存放起来，而当再度激活新的生命周期之后，马上回存这个对象机制。

① 重新改写 onRetainNonConfigurationInstance( ) 方法来传回想要保留的对象实体。

② 重新激活之后调用 getLastNonConfigurationInstance( ) 方法行回复原本被保留的物体。

## 13-4 资源内容的区域化

一般的开发方式，都是使用特定的语系进行程序设计，往往也因此而只能限定给特定的用户使用，无法扩大用户范围。这个章节就是介绍 Android 上的区域化程序设计方式。

## 13-4-1 支持的区域国别（地区）

以下列出Android2.2版本缩支持的国别（地区）语系（排列顺序依照Android开发者官方网站）。

- Chinese，PRC (zh_CN)：中国大陆。
- Chinese，Taiwan (zh_TW)：中国台湾地区。
- Czech (cs_CZ)：捷克。
- Dutch，Netherlands (nl_NL)：荷兰，荷兰文。
- Dutch，Belgium (nl_BE)：比利时，荷兰文。
- English，US (en_US)：美国，英文。
- English，Britain (en_GB)：英国，英文。
- English，Canada (en_CA)：加拿大，英文。
- English，Australia (en_AU)：澳大利亚，英文。
- English，New Zealand (en_NZ)：新西兰，英文。
- English，Singapore(en_SG)：新加坡，英文。
- French，France (fr_FR)：法国，法文。
- French，Belgium (fr_BE)：比利时，法文。
- French，Canada (fr_CA)：加拿大，法文。
- French，Switzerland (fr_CH)：瑞士，法文。
- German，Germany (de_DE)：德国，德文。
- German，Austria (de_AT)：澳大利亚，德文。
- German，Switzerland (de_CH)：瑞士，德文。
- German，Liechtenstein (de_LI)：列之敦斯登，德文。
- Italian，Italy (it_IT)：意大利，意大利文。
- Italian，Switzerland (it_CH)：瑞士，意大利文。
- Japanese (ja_JP)：日本。
- Korean (ko_KR)：韩国。
- Polish (pl_PL)：波兰。
- Russian (ru_RU)：俄罗斯。
- Spanish (es_ES)：西班牙。

对于用户而言，以上的国别（地区）语系可以通过Android手机上的"设定"→"语言与键盘"→"选取地区设定"来做适当的调整。

对于程序设计而言，如果没有特别的处理语系问题的程序，只需要将相关的资源按照项目预设的资源目录存储即可。可以在这些资源内容中直接使用中文或是其他语系的文字（别忘记资源目录就是在res/）。

然而，相关语系的资源不只是文字内容而已，还包括了图像、声音、影片等。举例来看，想处理一个操作界面，当按下按钮之后，会用语音汇报系统当时的状态，此时就可能会因为用户设定中文而发出国语语音；而设定英文的用户将会听到英语的说明。因此，就先从资源目录开始来介绍区域化的 Android 开发。

## ■ 13-4-2　进一步认识项目资源

　　Android 的项目资源形式可能是一般的文字（texts）、配置（layouts）、声音（sounds）、图像（graphics）及其他相关的静态数据项。任何一个 Android 项目中可以包含一或多个资源组（set），每组资源内容是用来应用在不同的装置组态之中，当用户执行了 Android 应用程序时，Android 会自动选择一组最佳最适合的资源组来应用，也就是说，Android 是自动选择，并不会每次执行 Android 应用程序都会询问用户，但是也不会对用户或是开发者造成困扰。举例而言，Android 设计者一开始就先处理预设的所有资源，接着继续开发中文语系，但是某些资源却没有另外再做处理，形成了预设资源比特定资源（目前范例是中文）还要更多的状况。而用户的 Android 装置上设定了日文语系，则将只会看到预设资源的内容，完全不会看到任何中文资源的内容；而设定为中文的用户，将可以看到所有的中文资源的内容，而对于预设资源中在中文资源所没有的内容，则只有该部分会以预设资源内容表现出来。这就是所谓 Android 的自动选择较佳的方式处理。

　　从上述说明中可以看到非常清楚的一点，就是预设资源的重要性是应该摆在第一优先处理。当然，这并不会对 Android 设计者造成任何困扰或是不便，因为一开始的设计方式就是如此。当一切设计就绪之后，再来好好将想要做特定资源处理的项目做语系上的转换即可（千万别想太多，通常是手动处理。当然也可以找到一些转换工具来使用，或是干脆自己写）。

　　因此，就像之前一样建立一个新的项目，也看到自动建立的资源目录，其结构如下：

　　在 LocalTest1 项目之下，存在了一个 res/ 资源目录。其子目录有 drawable-xxxx/、layout/ 及 values/3 个基本资源目录。现在就直接开始进行项目开发。首先处理 layout/main.xml，以配置出用户的操作界面。

res/layout/main.xml

```xml
<?xml version="1.0" encoding="utf-8"?>
<LinearLayout xmlns:android="http://schemas.android.com/apk/res/android"
    android:orientation="vertical"
    android:layout_width="fill_parent"
    android:layout_height="fill_parent"
    <Button
        android:id="@+id/click"
        android:layout_width="wrap_content"
        android:layout_height="wrap_content"
        android:text="@string/click"
        />
    <TextView
        android:id="@+id/tv"
        android:layout_width="fill_parent"
        android:layout_height="wrap_content"
        android:text="@string/hello"
        />
</LinearLayout>
```

因为使用了@string/click，并不存在默认的字符串资源中，所以接着继续处理 strings.xml：

① res/values/strings.xml

```xml
<?xml version="1.0" encoding="utf-8"?>
<resources>
    <string name="app_name">LocaleTest1</string>
    <string name="hello">Hello World, LocaleTest1!</string>
    <string name="click">Click Me</string>
</resources>
```

② src/tw.brad.android.test/LocaleTest1.java

```java
package tw.brad.android.test;

import android.app.Activity;
```

```java
import android.os.Bundle;
import android.view.View;
import android.view.View.OnClickListener;
import android.widget.Button;
import android.widget.TextView;
public class LocaleTest1 extends Activity implements OnClickListener {
    private Button click = null;
    private TextView tv = null;
    @Override
    public void onCreate(Bundle savedInstanceState) {
        super.onCreate(savedInstanceState);
        setContentView(R.layout.main);
        click = (Button)findViewById(R.id.click);
        tv = (TextView)findViewById(R.id.tv);
        click.setOnClickListener(this);
    }
    @Override
    public void onClick(View arg0) {
        tv.setText(R.string.mesg);
    }
}
```

设定当Button按下去之后，TextView的文字内容将会被修改成为R.string.mesg所代表的信息。此时因为尚未处理R.string.mesg，所以将会在该列的左方出现一个红色的叉。再度回到res/values/strings.xml补上该项资源内容。

res/values/strings.xml

```xml
<?xml version="1.0" encoding="utf-8"?>
<resources>
    <string name="app_name">LocaleTest1</string>
    <string name="hello">Hello World, LocaleTest1!</string>
    <string name="click">Click Me</string>
    <string name="mesg">Hello, Brad</string>
</resources>
```

到此为止应该可以正常运作如下：

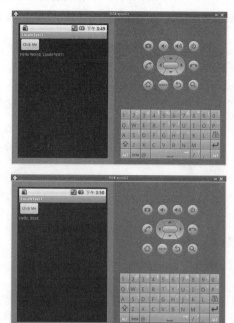

当按下 Click Me 的 Button 之后：

假设这一个庞大的 Android 项目开发已经开发完成。接着下来就是要处理区域化的事情……

在项目下新增一个 Android XML file：

按下Next之后，将会看到设定精灵的画面，只需要填入相关数据即可，如下：

上面的范例是建立出zh语系的资源文件，一切正常的状况下，按下Finish之后，会建立出res/values-zh的目录文件夹：

此时的重点就放在res/values-zh/strings.xml的文件内容：

```
<?xml version="1.0" encoding="utf-8"?>
<resources>
    <string name="app_name">区域化测试</string>
```

```
    <string name="hello">你好吗,全世界</string>
    <string name="click">按下去吧</string>
</resources>
```

程序都不需要做任何处理及修改,直接执行该项目的应用程序,首先看到图示的部分:

已经变成区域化测试的图标文字,接着点按进去。

来到这里都没有问题,全部内容都如愿地变成中文语系,当按下按钮之后:

这个信息是设定在res/values/stringx.xml中的：<string name="mesg"> Hello, Brad</string>。但是在res/values-zh/strings.xml中却没有处理到，因此，Android自动选择使用的较佳方式，就是呈现出预设资源中的内容，在这个项目中，还同时看到了两种不同的资源内容。

再来处理图标的资源。在菜单按下之后，会出现所有应用程序的图标及标题，这里的图示资源默认使用在res/drawable-xxxx/icon.png的图像文件。而使用区域化程序设计的处理手法类似前段说明的模式。直接手动新增一个res/下的子目录，名称为drawable-zh/，zh就是中文语系的代码。直接将所需要呈现的图标放在该子目录之下，并且覆盖掉原来的icon.png即可。处理方式相当简单。

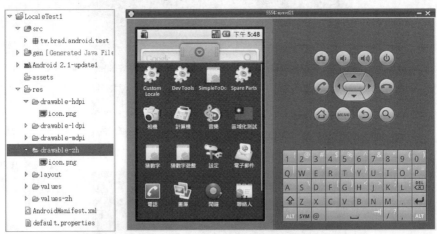

## 13-4-3　资源类型

资源类型大致上可以区分为以下几种：

（1）动画资源（Animation Resources）
- 事先制作完成的动画资源。
- 二维转换（twened animation）的动画资源，放在res/anim/目录下，使用R.anim方式存取。
- 逐一显示框架（frame by frame animation）的动画资源，放在res/drawable/目录架构下，使用R.drawable方式存取。

（2）颜色表资源（Color State List Resources）

颜色的对应资源，通常放在res/color/目录下，以R.color方式存取。

（3）绘图资源（Drawable Resources）

定义影像绘图的文件或是XML，通常放在res/drawable/目录下，以R.drawable方式进行存取。

（4）版面配置资源（Layout Resources）

应用程序的用户界面的版面配置资源，通常放在res/layout/目录下，以R.layout

方式进行存取。

（5）菜单资源（Menu Resources）

应用程序所需的菜单资源，通常放在 res/menu/ 目录下，以 R.menu 方式进行存取。

（6）字符串资源（String Resources）

应用程序中的显示字符串，通常放在 res/values/ 目录下，以 R.string，R.array 或是 R.plurals 方式进行存取。

（7）样式资源（Style Resources）

定义用户界面元素的呈现样式，通常放在 res/values/ 目录下，以 R.style 方式进行存取。

## ■ 13-4-4　区域化确认检查

这一节参考 Android 官方网站的确认检查表，笔者加上自身的经验提供给读者在进行区域化程序设计时参考使用。

（1）规划及设计检核表

• 确认区域化的策略。

• 哪些语系想要支持应用。

• 预设语系为什么。

• 哪些状况或是时机使用特定的语系或是默认语系。

• 确认应用程序中需要区域化的部分。

• 针对应用程序中的每个细节做分析，包括文字、影像、声音、音乐、数字、日期时间及货币的表示法。

• 用户完全看不到的地方就不需要处理区域化。

• 注意到应用程序的背景风格一定要搭配特定语系的风格。这样才有助于销售出应用程序。

• 程序代码的具体处理。

• 使用 R.string 或是 string.xml 方式替代掉直接写在程序代码中。

• 使用 R.drawable 或是 R.layout 方式替代掉直接写在程序代码中。

（2）内容检核表

• 建立完整的预设资源组。

• 确认转译的数据内容的正确性。

• 使用正确的语系文字内容的格式，尤其是像数字、货币及日期时间的表示格式。

• 有必要的话，不妨建立一套区域版本来控管。

（3）测试及发布检核表

• 测试应用程序所支持的所有语系，当然，可以的话，请该语系为母语的人来做测试是最好的方法。

- 测试预设资源组的正确性。
- 设定为不支持的语系,是否能正常呈现预设资源的内容。
- 测试支持语系中使用到预设资源内容的部分。
- 测试默认资源内容在装置的水平或是垂直的显示状况。
- 建立一个最终的发布版本。

# 14
Chapter

## 第14课　系统功能与装置控制

14-1　移动装置相关辨识

14-2　移动电话通话状态

14-3　移动电话用户相关数据

14-4　开发者基本道德

## 14-1 移动装置相关辨识

在一般的App设计中，很多时候会将移动装置的特性与一般PC产生不同的区隔，就是可以善加运用移动装置的特性，其中最基本的功能就是电话的使用。

Android.telephony.TelephonyManager对象实体就是用来获得用户移动装置上与电话相关的信息。取得该对象实体是通过调用getSystemService( )方法，并传入TELEPHONY_SERVICE参数即可。

声明变量：

```
private TelephonyManager tmgr;
```

取得对象实体：

```
tmgr = (TelephonyManager)getSystemService(TELEPHONY_SERVICE);
```

AndroidManifest.xml中需要开启相关权限：READ_PHONE_STATE。

```xml
<uses-permission android:name="android.permission.READ_PHONE_STATE"/>
```

接着下来就直接调用个别信息相关方法即可。

```java
package tw.brad.book.mydeviceinfo;

import android.app.Activity;
import android.os.Bundle;
import android.telephony.TelephonyManager;
import android.widget.TextView;

public class MainActivity extends Activity {
    private TextView info;
    private TelephonyManager tmgr;

    @Override
    protected void onCreate(Bundle savedInstanceState) {
        super.onCreate(savedInstanceState);
        setContentView(R.layout.activity_main);
        info = (TextView)findViewById(R.id.info);
        info.setText

        tmgr = (TelephonyManager)getSystemService(TELEPHONY_SERVICE);
```

```java
// 手机号码
String lineNumber = tmgr.getLine1Number( );
info.append("手机号码:" + lineNumber + "\n");

// IMEI码
String IMEI = tmgr.getDeviceId( );
info.append("IMEI:" + IMEI + "\n");

// IMSI码
String IMSI = tmgr.getSubscriberId( );
info.append("tmgr" + IMSI + "\n");

// 漫游状态
boolean isRoaming = tmgr.isNetworkRoaming( );
info.append("漫游:" + (isRoaming?"On":"Off") + "\n");

// 电信网络国别
String Country = tmgr.getNetworkCountryIso( );
info.append("电信网络国别:" + Country + "\n");

// 电信公司
String Operator = tmgr.getNetworkOperator( );
info.append("电信公司代号:" + Operator + "\"); 

// 电信公司名称
String OperatorName = tmgr.getNetworkOperatorName( );
info.append("电信公司名称:" + OperatorName + "\n");

// 移动网络类型
String[] NetworkTypes =
    {"UNKNOWN", "GPRS", "EDGE", "UMTS",
        "CDMA", "EVDO 0", "EVDO A", "1xRTT",
        "HSDPA", "HSUPA", "HSPA"};
String NetworkType = NetworkTypes[tmgr.getNetworkType( )];
info.append("移动网络类型:" + NetworkType + "\n");

// 移动通信类型
String[] PhoneTypeArray = {"NONE", "GSM", "CDMA"};
String PhoneType = PhoneTypeArray[tmgr.getPhoneType( )];
info.append("移动通信类型:" + PhoneType + "\n");
}

}
```

电话号码目前以笔者在台湾地区使用"中华电信"的状况下，是空字符串数据（不是null，null是没有插SIM卡的状况）。

IMEI（International Mobile Equipment Identity number）为国际移动装置识别符，也就是一般所谓的手机序号，算是移动装置的唯一识别码，共有15码。前6码为型号核准码，所以相同6码为相同机型；第78码为装配码，也代表产地。通常在移动装置上面输入 *#06# 就可以获得，其信息内容放在移动装置的内存中。

IMSI（International Mobile Subscriber Identity）为国际移动用户识别码，用来识别移动电话用户的唯一识别码，通常存放在SIM卡中。前3码为移动装置国码（台湾地区466）；第4、5码表示电信业者，没插SIM卡者没有数据。

如果App有绑定移动装置的特性，可以使用IMEI；如果有绑定电信业者或是与电话业务相关，可以使用IMSI。

实际测试结果：

## 14-2 移动电话通话状态

一样是通过TelephonyManager对象实体来处理，继承android.telephony.PhoneStateListener类别，改写onCallStateChanged( )方法。

再由TelephonyManager对象实体来调用listen( )方法，传入该对象实体，以及监听类型。

AndroidManifest.xml中需要开启相关使用权限：READ_PHONE_STATE。

```
<uses-permission android:name="android.permission.READ_PHONE_STATE"/>
```

以下以一个情境来处理，假设想要实现来电电话录音，通常会通过后台执行的Service来处理：

```
package tw.brad.book.mydeviceinfo;

import java.io.IOException;

import android.app.Service;
```

```java
import android.content.Intent;
import android.media.MediaRecorder;
import android.os.IBinder;
import android.telephony.PhoneStateListener;
import android.telephony.TelephonyManager;
import android.util.Log;
public class MyPhoneService extends Service {
    private TelephonyManager tmgr;
    private MediaRecorder mr;

    @Override
    public IBinder onBind(Intent arg0) {
        // TODO Auto-generated method stub
        return null;
    }

    @Override
    public void onCreate( ) {
        super.onCreate( );

        tmgr = (TelephonyManager)getSystemService(TELEPHONY_SERVICE);
        tmgr.listen(new MyPhoneStateListener( ),
PhoneStateListener.LISTEN_CALL_STATE);
    }

    private class MyPhoneStateListener extends PhoneStateListener {
      @Override
      public void onCallStateChanged(int state, String incomingNumber) {
            switch (state){
                case TelephonyManager.CALL_STATE_IDLE: // 闲置状态
                        if (mr != null){
                            // 结束电话录音
                            mr.stop( );
                            mr.release( );
                            mr = null;
                        }
                        break;
                case TelephonyManager.CALL_STATE_OFFHOOK:// 来电拿起话筒
                        // 开始电话录音
                        recordPhone( );
```

```
                    break;
                case TelephonyManager.CALL_STATE_RINGING:// 响铃中
                    // 显示对方来电
                    Log.i("brad", "RINGING:" + incomingNumber);
            }
        }
    }

    private void recordPhone( ){
        mr = new MediaRecorder( );
        mr.setAudioSource(MediaRecorder.AudioSource.MIC);
        mr.setOutputFormat(MediaRecorder.OutputFormat.MPEG_4);
        mr.setAudioEncoder(MediaRecorder.AudioEncoder.DEFAULT);
        mr.setOutputFile("/mnt/sdcard/brad1.mp4");

        try {
            mr.prepare( );
            mr.start( );
        } catch (IllegalStateException e) {
            // TODO Auto-generated catch block
            e.printStackTrace( );
        } catch (IOException e) {
            // TODO Auto-generated catch block
            e.printStackTrace( );
        }
    }
}
```

## 14-3 移动电话用户相关数据

在许多内建的 App 中会自动取得用户相关的设定数据，以方便用户存取服务，例如 Google Mail/Calendar，或是 Line 会自动取得用户的联系人数据等。以下简单介绍处理方式。

### ■ 14-3-1 用户账号

开启权限：

· ACCOUNT_MANAGER。

- GET_ACCOUNTS。

通过AccountManager对象实体来进行：

```
AccountManager mgr = (AccountManager)getSystemService(ACCOUNT_SERVICE);
```

接着调用getAccounts( )方法，传回Account[]数组。每一个Account[]数组元素中的Account对象，存在两个字符串属性name和type。

## ■ 14-3-2　取得联系人姓名

通过ContentProvider模式取得。传递第二个参数为搜寻关键词串，可以是联系人姓名或是电话号码的关键词，传回一个联系人姓名字符串数组。

```
//取得所有联系人姓名
static public String[] getContactsName(Context c, String key) {
    //取得内容解析器
    ContentResolver contentResolver = c.getContentResolver( );
    //设定你要从电话簿取出的字段
    String name = ContactsContract.CommonDataKinds.Phone.DISPLAY_NAME;
    String number = ContactsContract.CommonDataKinds.Phone.NUMBER;
    String[] projection = new String[]{      name, number};

    String selection = name + " like ? or " + number + " like ?";
    String[] selctionArgs = {"%" + key + "%", "%" + key + "%"};
    Cursor cursor;
    if (key != null){
        cursor = contentResolver.query(
                ContactsContract.CommonDataKinds.Phone.CONTENT_URI,
                projection, selection, selctionArgs,
                ContactsContract.CommonDataKinds.Phone.DISPLAY_NAME);
    }else{
        cursor = contentResolver.query(
                ContactsContract.CommonDataKinds.Phone.CONTENT_URI,
                projection, null, null,
                ContactsContract.CommonDataKinds.Phone.DISPLAY_NAME);
    }
    String[] contactsName = new String[cursor.getCount( )];
    String[] contactsPhone = new String[contactsName.length];

    String pretemp = ""; int ri = 0;
```

```
    for (int i = 0; i < cursor.getCount( ); i++) {
        //移到指定位置
        cursor.moveToPosition(i);
        if (!pretemp.equals(cursor.getString(1)) &&
                cursor.getString(1).trim( ).length( )>3){
            //取得第一个字段
            String phonenum = cursor.getString(1).replaceAll(" ", "");
            contactsName[ri] = cursor.getString(0) + "\n" + phonenum;
            contactsPhone[ri++] = phonenum;
            pretemp = cursor.getString(1);
        }
    }
    contactsName = Arrays.copyOf(contactsName, ri);
    contactsPhone = Arrays.copyOf(contactsPhone, ri);
    return contactsName;
}
```

### 14-3-3 用户的相簿

一般常看到相片管理的App调用使用。先取得ContentResolver对象实体，

```
ContentResolver resolver = getContentResolver( );
```

再来就进行query( )方法：

```
Cursor c = resolver.query(
    MediaStore.Images.Media.EXTERNAL_CONTENT_URI, null, null, null, null);
```

传回Cursor对象实体之后，就可以开始存取相片字段。

```
c.getString(c.getColumnIndexOrThrow(MediaStore.Images.Media.DATA));
```

回传相片文件在SD Card中实际文件位置。

## 14-4 开发者基本道德

本章节的应用其实对于用户而言，应该要尽告知之责。通常从Google play市场下载安装App，会有一个画面告知用户该App会开启的相关权限有哪些。但是大多

数的用户都是直接按下同意，急着看看是否安装完毕并马上使用。如果以这两小节所开启的权限，再加上一个因特网，App开发者可以非常轻松地将用户的电话信息、来电号码、秘密录音等数据传送到远程服务器。

"水能载舟亦能覆舟"，这类有关于用户隐私的相关数据，对于App开发者而言，可以有利于用户方便好用，却也能用在非正当的状态下。而这类的技术层面一点都不高，无奈目前笔者所接触到的用户，99.99%的比例都不仔细看看所下载的App所开启的权限，反正就是要玩要用。想一想，一个普通的游戏为何要读取用户账号、电话状态、通话记录……最可怕的就是因特网权限，可以将上述数据在后台中传送出去。

相信大多数App都是善良的，但是广告商的API要开启因特网权限才能轮播广告业主的广告内容，才能统计分析有效的营销等。

既然如此，大家就一起推广自由软件/开放原始码吧。

# MEMO

# 15
## Chapter

## 第15课　实际项目开发

15-1　弹指砖块王（Bricks Fighter）

15-2　掏金沙（Lode Runner）

15-3　炸弹超人（Bomb King）

15-4　其他应用程序开发项目

这只是笔者在"资策会"授课的开发范例项目，为了鼓励学员参加比赛并分享知相关参赛经验的实际案例，这是"中华电信"2012电信创新应用大赛中的移动应用/一般组优胜，笔者将此原始码开放，并进行开发说明，原始码公开的目的是希望借此提高读者学习动力。

http://innovation.hinet.net/telsoft/history/2012.html。

## 15-1 弹指砖块王（Bricks Fighter）

打砖块游戏已经在地球上存在30年以上，这个历久不衰的经典游戏移植到触控屏幕之后，如果只是传统的反弹棒的操控而已，那就只是一般的打砖块游戏。这款游戏利用触控屏幕的特性，可以瞬间画出一条实时反弹棒，玩家可以依照反弹的角度来自由画出反弹棒以反弹。而游戏的目的并非将所有的砖块都打光，是努力取得一开始出现在砖块最上方的宝物，依照关卡设计而有不同数量的宝物，玩家的策略在于取得所有的宝物。针对游戏的刺激性，玩家可以选择计时赛，各个关卡的计时不同。

对应上述的动机，将会应用以下技巧：
- 周期计时的线程。
- 自定义View类别成为游戏的主画面。
- 触控屏幕的手势侦测。
- 读取项目中的关卡数据。
- 动画效果。

## 15-1-1 App简易架构

## 15-1-2 欢迎页面

版面配置：res/layout/welcome.xml。

```xml
<?xml version="1.0" encoding="utf-8"?>
<RelativeLayout xmlns:android="http://schemas.android.com/apk/res/android"
    android:id="@+id/welcome"
    android:layout_width="fill_parent"
    android:layout_height="fill_parent"
    android:background="@drawable/bg0" >

    <ImageView
        android:id="@+id/logo"
        android:layout_width="wrap_content"
        android:layout_height="wrap_content"
        android:layout_centerInParent="true"
        android:src="@drawable/welcome" />

</RelativeLayout>
```

而为了增加效果，使用简单的渐进/增大/转动画处理：res/anim/a1.xml。

```xml
<?xml version="1.0" encoding="utf-8"?>
<set android:shareInterpolator="false"
    xmlns:android="http://schemas.android.com/apk/res/android"
    >
    <alpha
          android:fromAlpha="0.1"
          android:toAlpha="1.0"
          android:duration="2000"
          android:interpolator="@android:anim/accelerate_interpolator"
        />
    <scale
```

```xml
        android:fromXScale="0.0"
        android:fromYScale="0.0"
        android:toXScale="1.0"
        android:toYScale="1.0"
        android:pivotX="50%"
        android:pivotY="50%"
           android:duration="2000"
        />
    <rotate
        android:fromDegrees="359"
        android:toDegrees="0"
        android:pivotX="50%"
        android:pivotY="50%"
           android:duration="2000"
        />
</set>
```

开始编写 Activity：

```java
package tw.brad.android.games.BricksFighter;

import android.app.Activity;
import android.content.Intent;
import android.os.Bundle;
import android.view.View;
import android.view.View.OnClickListener;
import android.view.animation.Animation;
import android.view.animation.AnimationUtils;
import android.widget.ImageView;
import android.widget.RelativeLayout;
public class Welcome extends Activity {
    private ImageView logo;
    private RelativeLayout welcome;
    private Animation a1;

    @Override
    protected void onCreate(Bundle savedInstanceState) {
        super.onCreate(savedInstanceState);
        setContentView(R.layout.welcome);
```

```
        logo = (ImageView) findViewById(R.id.logo);
        welcome = (RelativeLayout) findViewById(R.id.welcome);
        a1 = AnimationUtils.loadAnimation(this, R.anim.a1);
        logo.startAnimation(a1);

        welcome.setOnClickListener(new OnClickListener( ) {
            @Override
            public void onClick(View v) {
                logo.clearAnimation( );
                Intent it = new Intent(Welcome.this, MainMenu.class);
                startActivity(it);
                Welcome.this.finish( );
            }
        });
        logo.setOnClickListener(new OnClickListener( ) {
            @Override
            public void onClick(View v) {
              logo.clearAnimation( );
              Intent it = new Intent(Welcome.this, MainMenu.class);
              startActivity(it);
              Welcome.this.finish( );
            }
        });
    }@Override
    public void finish( ) {
        super.finish( );
    }
}
```

玩家不会想看这个画面太久，但是开发者希望玩家可停留久一点以留下印象，所以经过权衡，当出现欢迎画面后，用户自行触摸屏幕离开，进入游戏关卡菜单。

## ■ 15-1-3 游戏关卡菜单

版面配置：res/layout/mainmenu.xml。

采用RelativeLayout。

```xml
<?xml version="1.0" encoding="utf-8"?>
<RelativeLayout xmlns:android="http://schemas.android.com/apk/res/android"
    android:id="@+id/mainmenu"
    android:layout_width="fill_parent"
    android:layout_height="fill_parent"
    android:background="@drawable/brickwall"
    >

    <LinearLayout
        android:id="@+id/mtitle"
        android:layout_width="wrap_content"
        android:layout_height="wrap_content"
        android:layout_alignParentTop="true"
        >
        <ImageView
            android:id="@+id/menu_title"
            android:layout_width="wrap_content"
            android:layout_height="wrap_content"
            android:src="@drawable/mlogo"
            />
    </LinearLayout>

    <LinearLayout
        android:id="@+id/func"
        android:layout_alignParentBottom="true"
        android:layout_width="fill_parent"
        android:layout_height="wrap_content"
        android:orientation="horizontal"
        android:gravity="center_horizontal|bottom"
        android:background="@drawable/menu_bottom"
        >
        <ImageView
            android:id="@+id/sound"
            android:src="@drawable/sound_on"
            android:layout_width="wrap_content"
            android:layout_height="wrap_content"
            android:layout_weight="1"
            />
```

```xml
        <ImageView
            android:id="@+id/limit"
            android:src="@drawable/time_nolimit"
            android:layout_width="wrap_content"
            android:layout_height="wrap_content"
            android:layout_weight="1"
            />
        <ImageView
            android:id="@+id/help"
            android:src="@drawable/help"
            android:layout_width="wrap_content"
            android:layout_height="wrap_content"
            android:layout_weight="1"
            />
        <ImageView
            android:id="@+id/exit"
            android:src="@drawable/exit"
            android:layout_width="wrap_content"
            android:layout_height="wrap_content"
            android:layout_weight="1"
            />
    </LinearLayout>
    <GridView
        android:id="@+id/gv"
        android:layout_below="@id/mtitle"
        android:layout_above="@id/func"
        android:layout_width="fill_parent"
        android:layout_height="fill_parent"
        android:numColumns="4"
        android:horizontalSpacing="32dp"
        android:layout_marginBottom="12dp"
        android:layout_marginTop="2dp"
        />
</RelativeLayout>
```

搭配个别影像文件后，呈现风格如下：

上方为标题列，下方为功能栏，中间全部空间为GridView的网格配置，其他部分就由Activity中处理。

以下就几个重点解说。

通过SharedPreferences取得用户游戏相关设定及状态信息。

```
sp = getSharedPreferences("gamedata", MODE_PRIVATE);
speditor = sp.edit( );

isSound = sp.getBoolean("sound", false);    // 是否播放音效
isLimit = sp.getBoolean("limit", false);    // 是否限时模式
user_level = sp.getInt("user_level", 0);    // 目前最高等级
```

用户按下下方Sound按钮的切换状态（限时模式也是相同处理手法）：

```
sound = (ImageView) findViewById(R.id.sound);
sound.setImageResource(resSound[isSound ? 1 : 0]);
sound.setOnClickListener(new OnClickListener( ) {
    public void onClick(View v) {
        isSound = isSound ? false : true;
        sound.setImageResource(resSound[isSound ? 1 : 0]);
        speditor.putBoolean("sound", isSound);
        speditor.commit( );
    }
});
```

将游戏选关处理搭配目前玩家的最高等级处理方式:

```java
private void reloadMenu( ) {
    String[] from = { "level_img", "level_txt" };
    int[] to = { R.id.level_img, R.id.level_txt };
    user_level = sp.getInt("user_level", 0);
    ArrayList<HashMap<String, Object>> item = new ArrayList<HashMap<String, Object>>( );
    for (int i = 0; i <= 27; i++) {
        // 以下正式发布用
        if (i <= user_level) {
            HashMap<String, Object> map1 = new HashMap<String, Object>( );
            map1.put("level_img", R.drawable.level_ed);
            map1.put("level_txt", i);
            item.add(map1);
        } else {
            HashMap<String, Object> map1 = new HashMap<String, Object>( );
            map1.put("level_img", R.drawable.level_yet);
            map1.put("level_txt", i);
            item.add(map1);
        }

        // 以下测试用
        // HashMap<String,Object> map1 = new HashMap<String,Object>( );
        // map1.put("level_img", R.drawable.level_ed);
        // map1.put("level_txt", i);
        // item.add(map1);
    }
    // 以下设定用户选按特定游戏关卡
    gv.setOnItemClickListener(new OnItemClickListener( ) {
        public void onItemClick(AdapterView<?> arg0, View arg1, int arg2,
                long arg3) {
            // 以下正式发布用
            if (arg2 <= user_level) {
                toGameView(arg2);
            }
```

```
                // 以下测试用
                // toGameView(arg2);
            }
        });
        gv.setSelection(user_level > 3 ? user_level - 4 : 0);
}
```

其中还分成正式发布和测试用,因为笔者可能会开发到很难的关卡,还要和一般玩家一样一路厮杀过关就太累了,所以测试用的部分可以直接玩任何一关。

上述的reloadMenu( )是在该Activity一开始的时候被调用,以及当玩家从游戏画面回到菜单的时候,也将会被调用,可能的差异在于是否已经玩到后面进阶的关卡。而最后调用的setSelection( )就是避免每次都是从第一关开始呈现,而玩家已经玩到后面关卡,所以直接定位菜单的一开始处在玩家最新关卡栏。

### 15-1-4 游戏主页

采用自定义View的方式来处理,主控的Activity为Main.java,负责处理:
- 载入游戏的初始画面。
- 游戏每一回合的结束,并呈现询问对话框。
- 监听游戏状态。

而游戏画面为PlayView.java中的自定义View类别,负责游戏过程中的所有程序。

因此先来检视最简单的版面配置(res/layout/main.xml):

```xml
<?xml version="1.0" encoding="utf-8"?>
<LinearLayout xmlns:android="http://schemas.android.com/apk/res/android"
    android:layout_width="fill_parent"
    android:layout_height="fill_parent"
    android:orientation="vertical"
    >
    <tw.brad.android.games.BricksFighter.PlayView
        android:id="@+id/playview"
        android:layout_width="fill_parent"
        android:layout_height="fill_parent"
        />
</LinearLayout>
```

这样就可以开始开发PlayView.java的游戏主页了。

主要有以下几个部分要进行绘制：
- 背景（特定数个影像随机出现，增加不同的感觉）。
- 上方游戏信息列。
- 砖块关卡。
- 中间为手势触控区，玩家可在此区域以手势画出自定义反弹棒。
- 反弹控制器（传统打砖块的反弹棒）。
- 最下列为左右移动反弹控制器。

加载游戏关卡，使用最简单的文字数据解析。

事先编写关卡数据放在raw/xxx，假设上图中的第七关为raw/level07，

```
35
0,6,6,6,5,6,6,6,0
0,5,7,5,7,5,7,5,0
6,5,0,6,5,6,0,5,6
6,9,0,6,6,6,0,9,6
6,0,0,0,0,0,0,0,6
6,8,6,6,0,6,6,8,6
5,6,5,6,5,6,5,6,5
```

第一列的35表示如果玩家玩限时模式，该关卡设定为35s。
之后的每一列表示砖块的一列，自定义如下。
- 0：空。
- 6：一般砖块（打击一次就破掉）。
- 5：坚硬砖块（打击三次才全破）。
- 7：炸弹（可以击破周围四周全部砖块）。

- 8：射击宝物（可以在一定时间内连续发射子弹）。
- 9：钢砖（打不破）。

定义全部关卡文件资源：

```
private static final int[] level = {
    R.raw.level00, R.raw.level01, R.raw.level02,
    R.raw.level03, R.raw.level04, R.raw.level05,
    R.raw.level06, R.raw.level07, R.raw.level08,
    R.raw.level09, R.raw.level10, R.raw.level11,
    R.raw.level12, R.raw.level13, R.raw.level14,
    R.raw.level15, R.raw.level16, R.raw.level17,
    R.raw.level18, R.raw.level19, R.raw.level20,
    R.raw.level21, R.raw.level22, R.raw.level23,
    R.raw.level24, R.raw.level25, R.raw.level26,
    R.raw.level27, R.raw.level28, R.raw.level29,
};
```

加载并同时解析关卡数据。

```
private void loadData( ){
    // 加载指定 level 的地图数据
    try {
        noAppleC.clear( );
        BufferedReader reader;
        reader =new BufferedReader(
        new InputStreamReader(res.openRawResource(level[play_level])));

        total_bricks = 0; mapLines = 0;
        map = new int[16][9];
        String temp1;
        String[] temp3;
        int i = 0, j = 0, v;

        temp1 = reader.readLine( );
        limitSeconds = Integer.parseInt(temp1);
        while ((temp1 = reader.readLine( ))!=null){
            temp3 = temp1.split(",");
            for(String temp4 : temp3){
                v = Integer.parseInt(temp4.trim( ));
```

```
                    if (v == 9) noAppleC.add(j);
                            // 有钢板的 column 不放宝物
                    map[i][j++] = v;
                    if (v>0 && v < 10) total_bricks++;
                }
                mapLines = i;
                i++; j=0;
            }
            reader.close( );
        } catch (IOException e) {
        }
    }
```

有了数据就可以开始处理影像图文件，在此笔者并不想要为不同尺寸屏幕而烦恼，所以影像资源会先配合玩家的移动装置的屏幕而进行比例计算而缩放。

先自定义一个init( )方法出来，当开启游戏画面时只需要执行一次的处理即可。

```
private void init( ){
    viewW = getWidth( );
    viewH = getHeight( );
    matrix.reset( );
    barW = viewW/5; barH = viewH/32;
    matrix.postScale((float)barW/bar.getWidth( ), (float)barH/bar.getHeight( ));
    bar = Bitmap.createBitmap(bar, 0, 0, bar.getWidth( ), bar.
    getHeight( ), matrix, false);
    matrix.reset( );
    matrix.postScale((float)barW/barfire.getWidth( ), (float)barH/barfire.
        getHeight( ));
    barfire = Bitmap.createBitmap(barfire, 0, 0, barfire.getWidth( ),
        barfire.getHeight( ), matrix, false);
    bottomY = viewH - viewH/6 + barH;
    matrix.reset( );
    matrix.postScale((float)viewW/bgW, (float)viewH/bgH);
    bg = Bitmap.createBitmap(bg, 0, 0, bgW, bgH, matrix, false);
    ...
    matrix.reset( );
```

```
        matrix.postScale((float)viewW/top.getWidth( ), (float)(viewH-
            bottomY)/top.getHeight( ));
        top = Bitmap.createBitmap(top, 0, 0, top.getWidth( ), top.
            getHeight( ), matrix, false);

        matrix.reset( );
        ballW = viewW/24; ballH = ballW;
        matrix.postScale((float)ballW/ball.getWidth( ), (float)ballH/ball.
            getHeight( ));
        ball = Bitmap.createBitmap(ball, 0, 0, ball.getWidth( ), ball.
            getHeight( ), matrix, false);

        matrix.reset( );
        appleW = barW/3; appleH = barH; appledy = appleH/4;
        matrix.postScale((float)appleW/apple.getWidth( ), (float)appleH/
            apple.getHeight( ));
        apple = Bitmap.createBitmap(apple, 0, 0, apple.getWidth( ), apple.
            getHeight( ), matrix, false);
    ...
    }
```

在进行缩放之前，先调用getWidth( )与getHeight( )方法取得玩家当前移动装置的实际宽高。

Matrix对象变量matrix是重要进行缩放的对象，从头到尾只用一个对象，所以要进行不同影像缩放之前要先调用reset( )方法。而影像尺存完全依照屏幕宽高的比例来处理，所以不需要为上千种不同的屏幕来烦恼。

准备就绪后，就来处理每次需要重新绘制的方法。

```
@Override
public void draw(Canvas canvas) {
    if (!isInit)
        init( );

    canvas.drawBitmap(bg, 0, 0, p);
    canvas.drawBitmap(arrow_bg, 0, bottomY, p);
    canvas.drawBitmap(top, 0, 0, p2);

    if (!isStart) {
        canvas.drawBitmap(hand, (viewW - ctW) / 2, barY - ctH - barH
```

```
                    * 2, p);
    }
// 限时模式的倒数计时
    if (isLimit) {
        switch (limitSeconds - intTimer / 100) {
        case -1:
            stopRound( );
            return;
        case 1:
            canvas.drawBitmap(c01, (viewW - ctW) / 2,
                barY - ctH - barH * 2, p);
            break;
        case 2:
            canvas.drawBitmap(c02, (viewW - ctW) / 2,
                barY - ctH - barH * 2, p);
            break;
        case 3:
            canvas.drawBitmap(c03, (viewW - ctW) / 2,
                barY - ctH - barH * 2, p);
            break;
        case 4:
            canvas.drawBitmap(c04, (viewW - ctW) / 2,
                barY - ctH - barH * 2, p);
            break;
        case 5:
            canvas.drawBitmap(c05, (viewW - ctW) / 2,
                barY - ctH - barH * 2, p);
            break;
        case 6:
            canvas.drawBitmap(c06, (viewW - ctW) / 2,
                barY - ctH - barH * 2, p);
            break;
        case 7:
            canvas.drawBitmap(c07, (viewW - ctW) / 2,
                barY - ctH - barH * 2, p);
            break;
```

```java
        case 8:
            canvas.drawBitmap(c08, (viewW - ctW) / 2,
                    barY - ctH - barH * 2, p);
            break;
        case 9:
            canvas.drawBitmap(c09, (viewW - ctW) / 2,
                    barY - ctH - barH * 2, p);
            break;
        case 10:
            canvas.drawBitmap(c10, (viewW - ctW) / 2,
                    barY - ctH - barH * 2, p);
            break;
    }
}
float ballCX = ballX + ballW / 2, ballCY = ballY + ballH / 2;
float ballX2 = ballX + ballW, ballY2 = ballY + ballH; // 绘制目前的砖块区
for (int r = 0; r < map.length; r++) {
    for (int c = 0; c < map[r].length; c++) {
        if (map[r][c] > 0) {
            float bx = c * (brickW), by = (r + brickStartR)
                    * (brickH + 1);
            canvas.drawBitmap(bks[map[r][c]], bx, by, null);
        } else if (map[r][c] < 0) {
            float bx = c * (brickW), by = (r + brickStartR)
                    * (brickH + 1);
            switch (map[r][c]) {
                case -7:
                    canvas.drawBitmap(bomb1, bx, by, null);
                    map[r][c]++;
                    break;
                case -6:
                    canvas.drawBitmap(bomb2, bx, by, null);
                    map[r][c]++;
                    break;
                case -5:
                    canvas.drawBitmap(bomb3, bx, by, null);
```

```
                    map[r][c]++;
                    break;
                case -4:
                    canvas.drawBitmap(bomb4, bx, by, null);
                    map[r][c]++;
                    break;
                case -3:
                    canvas.drawBitmap(bomb3, bx, by, null);
                    map[r][c]++;
                    break;
                case -2:
                    canvas.drawBitmap(bomb2, bx, by, null);
                    map[r][c]++;
                    break;
                case -1:
                    canvas.drawBitmap(bomb1, bx, by, null);
                    map[r][c]++;
                    break;
                }
            }
        }
    }
    for (int i = 0; i < appleTask.length; i++) {
        if (appleTask[i] != null) {
            canvas.drawBitmap(apple, appleTask[i].getX( ),
                appleTask[i].getY( ), null);
        }
    }
// 发射子弹
    synchronized (bullets) {
        for (Bullet bt : bullets) {
            if (!bt.isHit) {
                canvas.drawBitmap(bullet, bt.bx, bt.by, null);
            }
        }
    }
```

```java
        canvas.drawBitmap(ball, ballX, ballY, null);
        canvas.drawBitmap(isFireMode ? barfire : bar, barX, barY, null);
        // 绘制玩家的手势反弹棒
        if (isShowDrawLine
            && System.currentTimeMillis() - startShowDrawTime <= displayTime) {
                canvas.drawBitmap(drawbar, drawbarX, drawbarY, null);
        } else {
                isShowDrawLine = false;
        }
        // 绘制上方信息列
        canvas.drawText(strTimer, viewW - 100, 32, timerPaint);
        canvas.drawText(res.getString(R.string.stage) + play_level, 10, 32,
                stagePaint);
        if (isLimit) {
                canvas.drawText(res.getString(R.string.timelimit) + " "
                        + limitSeconds + "s", viewW / 3, 32, limitPaint);
        }
    // 过关或是失败的图案呈现
        if (isStopRound) {
                if (isWin) {
                        canvas.drawBitmap(success, 0, 0, null);
                } else {
                        canvas.drawBitmap(fail, 0, 0, null);
                }
        }
}
```

处理绘制的大原则就是不要有太多或是太复杂的处理程序，尽量单纯到只是做绘制的工作而已。相关计算控制变化都应该在各自的对象线程中处理，务必以使时间周期绘制工作简单为原则。

最后处理各自对象的线程，其实也只有两种对象：反弹球和子弹。

而碰撞侦测的原则是：

- 移动对象碰撞静止对象，是由移动对象进行侦测。
- 移动速度快与移动速度慢碰撞，通常是由移动速度快者进行判断。
- 移动速度一样者，就任选一个即可。

最后，呈现成功过关画面给读者：

## 15-2 掏金沙（Lode Runner）

### ■ 15-2-1 开发动机

　　Lode Runner（中文名称为"掏金沙"）是一款相当经典的游戏，笔者首次是在1985年的Apple Ⅱ上玩的PC游戏，当年的架构画面非常简单。游戏的主轴是一位主角，在每一个关卡中找寻宝物闯关，但是会有敌人前来阻挠，游戏主角可以运用智能避开，或是挖开脚下的砖块使敌人掉下去，暂时躲过一劫。下图为当年Apple Ⅱ上的屏幕截图（http://en.wikipedia.org/wiki/File:Lode_Runner.jpg）。

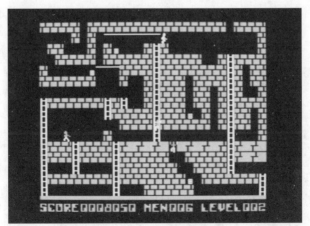

　　因为其游戏主题中包含了智能及耐心的游戏特性，加上关卡的丰富变化性，也曾在红白机上改编为"掏金沙"。

原游戏关卡主要由砖块、硬砖、楼梯和单杠构成，而宝物出现的位置在每关卡各有不同的设计，敌人的初始位置与数量也会随关卡而有不同的规划。笔者对于这款游戏非常怀念，但是在2010年初努力搜寻各个Market，就是无法找到原汁原味的Android Game，而其他类似的游戏，不是太过单调，关卡设计不多，操作不易，就是以3D呈现而失去了原来的智慧性。因此决定自己下手来设计，尽量呈现出原来的感觉，关卡数150关以上，而操作模式务必发挥触控屏幕的特性。终于花了约两个月的时间完成这个项目。以下就以该项目开放原始码方式，为读者简单介绍。

### ■ 15-2-2　着手规划

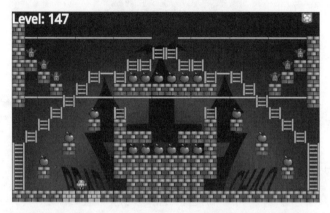

这是第147关的初始画面。下方的Android是游戏主角，蓝色的坏鸟（Blue Bad Bird，BB Bird）就是敌人，而苹果（或是可能出现芒果）就是宝物，屏幕的右上方就是当吃完所有苹果后，要努力抵达的胜利徽章，灰色砖块的显示，表示当Android停下来的时候可以挖掘的砖块。

为了达到持续开发不同的关卡，还特地开发出关卡编辑器，也是利用触控屏幕的特性，用手指就可以轻易设计出不同特色的关卡。最后还将自编关卡的功能提供给玩家使用，享受自导自演的乐趣。

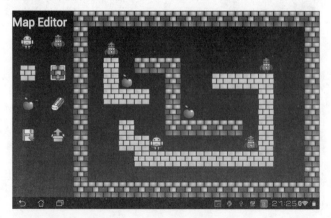

## 15-2-3 游戏架构

大致上与上一个章节的Bricks Fighter类似，所以直接以游戏画面说明较为易懂。

欢迎画面用几张图叙述游戏的故事性。在Android农庄里，种了许多苹果树和芒果树，Android农夫正在开心地采收，而BB Bird城堡的人却在对岸觊觎着这些丰收的水果。

有一天早上，Android农夫发现苹果和芒果少了许多。

于是决定在晚上偷偷观察，终于知道是隔壁BB Bird城堡的坏蛋来偷的。

决定派出Android Runner出动前往BB Bird城堡（就是各式各样的关卡）将水果拿回来。

这些图片的处理，事实上只是一个Activity中的一个ImageView，当用户触摸屏幕后切换到下一张图像文件，如果是触摸"Skip"文字字样，则离开Activity，进入到关卡菜单。

res/layout/welcome.xml

```xml
<?xml version="1.0" encoding="utf-8"?>
<RelativeLayout xmlns:android="http://schemas.android.com/apk/res/android"
    android:layout_width="fill_parent"
    android:layout_height="fill_parent"
    android:background="@drawable/bg_welcome"
    >
    <ImageView
      android:id="@+id/welcome_img"
        android:layout_width="fill_parent"
        android:layout_height="fill_parent"
        android:src="@drawable/logo"
    />
    <TextView
      android:id="@+id/welcome_skip"
        android:layout_width="wrap_content"
        android:layout_height="wrap_content"
        android:text="@string/welcome_skip"
        android:textSize="36dp"
        android:textStyle="bold|italic"
        android:textColor="#ffffff00"
```

```xml
            android:layout_alignParentBottom="true"
            android:layout_alignParentRight="true"
            android:layout_marginRight="42dp"
            />
</RelativeLayout>
```

而在主程序中的处理就非常简单。

Welcome.java

```java
package tw.brad.apps.LodeRunner;

import android.app.Activity;
import android.content.Intent;
import android.os.Bundle;
import android.view.View;
import android.view.View.OnClickListener;
import android.view.WindowManager;
import android.view.animation.Animation;
import android.view.animation.AnimationUtils;
import android.widget.ImageView;
import android.widget.TextView;

public class private Welcome extends Activity {
    private ImageView welcome_img;
    private TextView welcome_skip;
    private int nowImgIndex = 0;
    private int[] imgs = {
              story2, story4
            R.drawable.story1, R.drawable.story2,
            R.drawable.story3, R.drawable.story4
    };

    @Override
    public void onCreate(Bundle savedInstanceState) {
        super.onCreate(savedInstanceState);
            // 防止进入休眠
        getWindow( ).setFlags(WindowManager.LayoutParams.FLAG_KEEP_
            SCREEN_ON,
```

```java
            WindowManager.LayoutParams.FLAG_KEEP_SCREEN_ON);
    setContentView(R.layout.welcome);

    welcome_img = (ImageView)findViewById(R.id.welcome_img);
    Animation an = AnimationUtils.loadAnimation(this, R.anim.myanim);
    welcome_img.startAnimation(an);

    welcome_skip = (TextView)findViewById(R.id.welcome_skip);
    welcome_skip.setOnClickListener(new OnClickListener( ){
        @Override
        public void onClick(View v) {
            // 进入加载数据画面
            Intent intent =
                    new Intent(Welcome.this, MainMenu.class);
            startActivity(intent);
            Welcome.this.finish( );
        }
    });
    welcome_img.setOnClickListener(new OnClickListener( ) {
        @Override
        public void onClick(View arg0) {
            if (nowImgIndex == 4){
                // 进入加载菜单数据画面
                Intent intent =
                    new Intent(Welcome.this, MainMenu.class);
                startActivity(intent);
                Welcome.this.finish( );
            }else {
    welcome_img.setImageResource(imgs[nowImgIndex++]);
            }
        }
    });
}
}
```

### 15-2-4 关卡菜单

关卡菜单的处理手法与 Bricks Fighter 可以说是完全相同。

## 15-2-5　游戏画面

游戏画面完全是以利用自定义View的开发方式处理。重点如下：

- 整个屏幕不分尺寸大小，规划成16×24的基本单位来处理。
- 关卡资源文件以int[]列存放，仅需要才作加载，因此不需过久的加载等候时间。
- 依照玩家自行设定，可以调整游戏执行快慢，重点是以智能过关，不是速度快慢过关。

一开始处理所有影像元素，自行开发一个自定义方法处理：

```
private Bitmap fitBitmap(int resId){
    Bitmap temp = BitmapFactory.decodeResource(res, resId);
    int bmp_bw = temp.getWidth( );
    int bmp_bh = temp.getHeight( );
    matrix.reset( );
    matrix.postScale((wp)/bmp_bw, (hp)/bmp_bh);
    return Bitmap.createBitmap(temp, 0, 0, bmp_bw, bmp_bh, matrix, true);
}
```

所以每个影像元素都是一个基本单位大小。

当使用过关而进到下一关，差异在于关卡数据读取，以及各项数据的初始化处理：

```
private void restartGame(int n){
    if (isSound) {
        Music.stop(this);
        if (bgmusic != null){
            bgmusic.setLooping(true);
            bgmusic.setScreenOnWhilePlaying(false);
            bgmusic.setVolume(0.2f,0.2f);
```

```
                bgmusic.start( );
            }
        }
        System.gc( );
        try{
                gview = new GameView(this, null, loadGameData(n));
                // 处理游戏背景
                gview.setBackgroundResource(R.drawable.bg);
                setContentView(gview);
        }catch(Exception e){
        }
}
```

加载关卡方法与 Bricks Fighter 类似，差异只是游戏元素的定义而已。

```
private int loadGameData(int n) throws Exception {
    int eCount = 0;
    // 加载指定 level 的地图数据
    BufferedReader reader;
    if (n == -1) {
            reader = new BufferedReader(new FileReader(play_userfile));}
    else {
            reader = new BufferedReader(new InputStreamReader(
                res.openRawResource(leveldata[n])));
    }
    map = new int[16][24];
    String temp1;
    String[] temp3;
    int i = 0, j = 0, v;
    while ((temp1 = reader.readLine( )) != null) {
        temp3 = temp1.split(",");
        for (String temp4 : temp3) {
            v = Integer.parseInt(temp4);
            map[i][j++] = v;
            if (v == 7)
                    eCount++;
        }
```

```
                i++;
                j = 0;
        }
        reader.close( );
        return eCount;
}
```

传回值为这关的宝物个数，以利计算是否已经拿完该关出现的所有宝物，才会出现隐形楼梯抵达胜利徽章。

而关卡文件类似如下内容（res/raw/level037）。

```
0,0,0,0,0,0,0,0,0,0,0,0,0,0,0,0,0,0,0,0,0,0,0,-9
0,0,3,3,3,3,3,3,3,3,3,3,3,3,3,3,3,3,3,3,3,-1,1
2,1,0,0,0,0,1,0,0,0,0,0,0,0,0,0,0,1,0,0,0,1,0
2,0,1,1,1,1,0,0,0,0,0,0,0,0,0,0,0,0,1,1,1,1,0
2,0,1,2,9,0,0,0,0,0,0,0,0,0,0,0,7,0,9,2,1,0
2,0,0,2,1,1,1,1,1,1,1,1,1,1,1,1,1,1,1,2,1,0
2,0,1,9,0,1,1,1,1,9,0,1,0,9,1,1,1,1,1,9,1,0
2,0,0,1,1,1,1,1,1,1,1,1,1,1,1,1,1,1,1,2,1,0
2,0,1,1,1,1,1,1,1,1,1,1,1,1,1,1,1,1,1,2,1,0
2,0,1,1,1,1,1,1,1,0,0,0,0,0,1,1,1,1,1,2,1,0
2,0,1,1,1,1,1,1,1,0,0,0,0,2,1,1,1,1,1,2,1,0
2,0,1,1,1,1,1,1,1,9,0,0,0,2,1,1,1,1,1,2,1,0
2,0,0,0,0,0,0,0,0,0,1,1,1,1,1,2,3,3,3,3,2,0,0
2,0,0,0,0,0,0,0,0,1,1,1,1,1,0,2,0,0,0,0,0,0,0
2,0,0,0,0,8,0,0,0,0,1,1,1,1,1,0,0,0,7,0,0,0,0,0
1,1,1,1,1,1,1,1,1,1,1,1,1,1,1,1,1,1,1,1,1,1,1
```

表现出如下图所示。

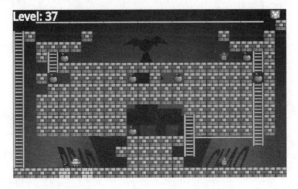

## 15-2-6 关卡地图

一开始真的是以手工编写数字数据的关卡，当写到第10关之后，已经无法再继续下去了，而当初立下志愿是至少150关，于是马上着手撰写关卡编辑器。

没有XML的版面处理，直接开发在Activity中，整个关卡编辑器就是这个Activity而已，一点也不复杂。

重点特色说明：
- 左侧是目前准备绘制的影像元素，会出现"+"字号。
- 利用触控屏幕特性直接在屏幕点触或是滑动绘制。
- Android只有一个，所以依照最新点触位置为主。
- 一边绘制可以一边思考关卡特性。

RunnerMapEditorActivity.java

```
package tw.brad.android.games.RunnerMapEditor;

import java.io.BufferedReader;
import java.io.BufferedWriter;
import java.io.FileNotFoundException;
import java.io.FileReader;
import java.io.FileWriter;
import java.io.IOException;

import android.app.Activity;
import android.app.AlertDialog;
import android.content.Context;
import android.content.DialogInterface;
import android.content.SharedPreferences;
import android.content.res.Resources;
import android.graphics.Bitmap;
import android.graphics.BitmapFactory;
import android.graphics.Canvas;
import android.graphics.Color;
import android.graphics.Matrix;
import android.graphics.Paint;
import android.graphics.Paint.Style;
import android.os.Bundle;
import android.util.AttributeSet;
import android.util.DisplayMetrics;
import android.util.Log;
```

```java
import android.view.LayoutInflater;
import android.view.Menu;
import android.view.MenuInflater;
import android.view.MenuItem;
import android.view.MotionEvent;
import android.view.View;
import android.widget.EditText;
import android.widget.Toast;
public class RunnerMapEditorActivity extends Activity {
    private MapEditor mapeditor;
    private Resources res;
    private int sw, sh;
    private float wp, hp;
    private String savename, openname;
    private SharedPreferences sp;
    private SharedPreferences.Editor sp_editor;
    private int now_level;
    private boolean isNew;

    @Override
    public void onCreate(Bundle savedInstanceState) {
        super.onCreate(savedInstanceState);

        res = getResources( );
        DisplayMetrics dm = res.getDisplayMetrics( );
        sw = dm.widthPixels; sh = dm.heightPixels;
        wp = sw / 24.0f; hp = sh / 16.0f;
        sp = getSharedPreferences("level", MODE_PRIVATE);
        sp_editor = sp.edit( );

        mapeditor = new MapEditor(this, null);
        setContentView(mapeditor);

        now_level = -1; isNew = true;
    }

  @Override
  public boolean onCreateOptionsMenu(Menu menu) {
        MenuInflater inflater = getMenuInflater( );
        inflater.inflate(R.menu.optionmenu, menu);
```

```java
            return super.onCreateOptionsMenu(menu);
    }

@Override
public boolean onOptionsItemSelected(MenuItem item) {
        switch (item.getItemId( )){
            case R.id.save:
                    mapeditor.saveMap( );
                    break;
            case R.id.open:
                    mapeditor.openMap( );
                    break;
            case R.id.clear:
                    mapeditor.clearMap( );
                    break;
            case R.id.newlevel:
                    now_level = -1; isNew = true;
                    mapeditor.clearMap( );
                    break;
        }
        return super.onOptionsItemSelected(item);
}

    private class MapEditor extends View {
     private int[][] map = new int[16][24];
            private Bitmap bmp_b, bmp_1, bmp_11, bmp_h, bmp_a, bmp_man,
            bmp_enemy, bmp_next;
            private int[] mode = {0,1,2,3,7,8,9,-1};
            private int now = 1;
            private int now_r, now_c;
            private Paint info_txt;
            private int count_man = 0;

            public MapEditor(Context context, AttributeSet attrs) {
                    super(context, attrs);

            bmp_b = BitmapFactory.decodeResource(res, R.drawable.bricks);
                    int bmp_bw = bmp_b.getWidth( );
                    int bmp_bh = bmp_b.getHeight( );

                    Matrix matrix = new Matrix( );
```

```
matrix.postScale((wp)/bmp_bw, (hp)/bmp_bh);
bmp_b = Bitmap.createBitmap(bmp_b, 0, 0, bmp_bw, bmp_
bh, matrix, true);

bmp_l = BitmapFactory.decodeResource(res, R.drawable.ladder);
bmp_bw = bmp_l.getWidth( );
bmp_bh = bmp_l.getHeight( );
matrix.reset( );
matrix.postScale((wp)/bmp_bw, (hp)/bmp_bh);
bmp_l = Bitmap.createBitmap(bmp_l, 0, 0, bmp_bw, bmp_
    bh, matrix, true);

bmp_l1 = BitmapFactory.decodeResource(res, R.drawable.
ladder_1);
bmp_bw = bmp_l1.getWidth( );
bmp_bh = bmp_l1.getHeight( );
matrix.reset( );
matrix.postScale((wp)/bmp_bw, (hp)/bmp_bh);
bmp_l1 = Bitmap.createBitmap(bmp_l1, 0, 0, bmp_bw, bmp_
bh, matrix, true);

bmp_h = BitmapFactory.decodeResource(res, R.drawable.
hbar);
bmp_bw = bmp_h.getWidth( );
bmp_bh = bmp_h.getHeight( );
matrix.reset( );
matrix.postScale((wp)/bmp_bw, (hp)/bmp_bh);
bmp_h = Bitmap.createBitmap(bmp_h, 0, 0, bmp_bw, bmp_
bh, matrix, true);

bmp_a = BitmapFactory.decodeResource(res, R.drawable.
apple);
bmp_bw = bmp_a.getWidth( );
bmp_bh = bmp_a.getHeight( );
matrix.reset( );
matrix.postScale((wp)/bmp_bw, (hp)/bmp_bh);
bmp_a = Bitmap.createBitmap(bmp_a, 0, 0, bmp_bw, bmp_
bh, matrix, true);

bmp_man = BitmapFactory.decodeResource(res, R.drawable.
    android);
```

```java
bmp_bw = bmp_man.getWidth( );
bmp_bh = bmp_man.getHeight( );
matrix.reset( );
matrix.postScale((wp)/bmp_bw, (hp)/bmp_bh);
bmp_man = Bitmap.createBitmap(bmp_man, 0, 0, bmp_bw,
bmp_bh, matrix, true);

bmp_enemy = BitmapFactory.decodeResource(res,
R.drawable.enemy);
bmp_bw = bmp_enemy.getWidth( );
bmp_bh = bmp_enemy.getHeight( );
matrix.reset( );
matrix.postScale((wp)/bmp_bw, (hp)/bmp_bh);
bmp_enemy = Bitmap.createBitmap(bmp_enemy, 0, 0, bmp_
bw, bmp_bh, matrix, true);

bmp_next = BitmapFactory.decodeResource(res, R.drawable.next);
bmp_bw = bmp_next.getWidth( );
bmp_bh = bmp_next.getHeight( );
matrix.reset( );
matrix.postScale((wp)/bmp_bw, (hp)/bmp_bh);
bmp_next = Bitmap.createBitmap(bmp_next, 0, 0, bmp_bw,
bmp_bh, matrix, true);

now_r = now_c = 0;

info_txt = new Paint( );
info_txt.setColor(Color.WHITE);
info_txt.setTextSize(36);
info_txt.setStyle(Style.FILL_AND_STROKE);
info_txt.setStrokeWidth(2);

    clearMap( );
}

@Override
public boolean onTouchEvent(MotionEvent event) {
    int click_r = (int)(event.getY( )/hp);
    int click_c = (int)(event.getX( )/wp);
    if (click_r == now_r && click_c == now_c){
        now = (now == mode.length-1)?0:now+1;
    }else {
```

```java
                now_r = click_r; now_c = click_c;
        }
        if (count_man == 1 && mode[now] == 8)now = (now == mode.length-1)?0:now+1;
        if (count_man == 0 || (count_man ==1 && mode[now] != 8)){
                map[click_r][click_c] = mode[now];
                postInvalidate( );
        }
        return super.onTouchEvent(event);
}

@Override
protected void onDraw(Canvas canvas) {
    super.onDraw(canvas);
    canvas.drawText("Level: " + now_level, 4, 40, info_txt);
        count_man = 0;
        for (int r = 0; r<map.length; r++){
                for (int c = 0; c<map[r].length; c++){
                    if (map[r][c]==1){
                    canvas.drawBitmap(bmp_b, c*wp, r*hp, null);
                    }else if (map[r][c]==2){
                    canvas.drawBitmap(bmp_l, c*wp, r*hp, null);
                    }else if (map[r][c]==3){
                    canvas.drawBitmap(bmp_h, c*wp, r*hp, null);
                    }else if (map[r][c]==7){   // 敌人
                    canvas.drawBitmap(bmp_enemy, c*wp, r*hp, null);
                    }else if (map[r][c]==8){   // Android
                    canvas.drawBitmap(bmp_man, c*wp, r*hp, null);
                        count_man = 1;
                    }else if (map[r][c]==9){   // Apple
                    canvas.drawBitmap(bmp_a, c*wp, r*hp, null);
                    }else if (map[r][c]==-9){  // 过关
                    canvas.drawBitmap(bmp_next, c*wp, r*hp, null);
                    }else if (map[r][c]==-1){  // 过关梯
                    canvas.drawBitmap(bmp_ll, c*wp, r*hp, null);
                    }
                }
        }
    }
}
```

```java
private void clearMap( ){
    map = new int[16][24];
    for (int j=0; j<24; j++){
        map[15][j] = 1;
    }
    map[0][23] = -9;
    postInvalidate( );
}
private void saveMap( ){
    LayoutInflater dialog = LayoutInflater.
        from(RunnerMapEditorActivity.this);
    final View dview = dialog.inflate(R.layout.dialog_save, null);
    EditText et = (EditText)dview.findViewById(R.id.savename);
    et.setText(sp.getInt("level", 0) + "");

    AlertDialog.Builder builder = new AlertDialog.
    Builder(RunnerMapEditorActivity.this);

    builder.setTitle("Save Level");
    builder.setView(dview);
    builder.setCancelable(true);
    builder.setPositiveButton("OK", new DialogInterface.
    OnClickListener( ){
        @Override
        public void onClick(DialogInterface arg0, int arg1) {
            EditText et = (EditText)dview.findViewById(R.
                id.savename);

            savename = et.getText( ).toString( );
            String header = savename;
            if (savename.length( )==1){
                savename = "00" + savename;
            }else if (savename.length( )==2){
                savename = "0" + savename;
            }

            try {
                BufferedWriter bw = new BufferedWriter(
                    new FileWriter( "/mnt/sdcard/level" + savename));

                StringBuffer line;
```

```java
                    for (int r=0; r<map.length; r++){
                        line = new StringBuffer( );
//                      line.append(header + ":");
                        for (int c=0; c<map[r].length; c++){
                            line.append(map[r][c]);
                            if (c==map[r].length-1){
                                line.append("\n");
                            }else{
                                line.append(",");
                            }
                        }
                        bw.write(line.toString( ));
                    }
                    bw.flush( );
                    bw.close( );
                    if (isNew){
                        sp_editor.putInt("level", Integer.
                            parseInt(savename)+1);
                        sp_editor.commit( );
                    }
                    Toast.makeText(RunnerMapEditorActivity.this,
                        "Save OK", Toast.LENGTH_LONG).show( );
                } catch (FileNotFoundException e) {
                    // TODO Auto-generated catch block
                    e.printStackTrace( );
                } catch (IOException e) {
                    // TODO Auto-generated catch block
                    e.printStackTrace( );
                }
            }
        });
        AlertDialog alert = builder.create( );
        alert.show( );
    }
    private void openMap( ){
        LayoutInflater dialog = LayoutInflater.from
            (RunnerMapEditorActivity.this);
        final View dview = dialog.inflate(R.layout.dialog_open, null);
```

```java
EditText et = (EditText)dview.findViewById(R.id.openname);
AlertDialog.Builder builder = new AlertDialog.Builder
        (RunnerMapEditorActivity.this);

builder.setTitle("Open Level");
builder.setView(dview);
builder.setCancelable(true);
builder.setPositiveButton("OK", new DialogInterface.
        OnClickListener( ){
    @Override
    public void onClick(DialogInterface arg0, int arg1) {
        EditText et = (EditText)dview.findViewById(R.
            id.openname);

        openname = et.getText( ).toString( );
  now_level = Integer.parseInt(openname); isNew = false;
        if (openname.length( )==1){
            openname = "00" + openname;
        }else if (openname.length( )==2){
            openname = "0" + openname;
        }
        try {
            BufferedReader reader = new BufferedReader(
                new FileReader("/mnt/sdcard/level" +
                    openname));
            String temp1;
            String[] temp3;
            int v, i = 0, j = 0;
            while ((temp1 = reader.readLine( ))!=null){
                temp3 = temp1.split(",");
                for(String temp4 : temp3){
                    v = Integer.parseInt(temp4);
                    map[i][j++] = v;
                }
                i++; j=0;
            }
            reader.close( );
            postInvalidate( );
            Toast.makeText(RunnerMapEditorActivity.
```

```
                       this, "Load OK", Toast.LENGTH_LONG).show( );
                } catch (FileNotFoundException e) {
                       now_level = -1; isNew = true;
                       e.printStackTrace( );
                } catch (IOException e) {
                       // TODO Auto-generated catch block
                       e.printStackTrace( );
                }
            }
        });
        AlertDialog alert = builder.create( );
        alert.show( );
    }
}
```

### 15-2-7 敏感争议

比起经典的Lode Runner还多带了一个明确的故事性，但是也利用故事隐含了讽刺性。苹果和芒果刚好是另外两个阵营的代表性水果，游戏主轴不是掏金沙，而改成吃掉苹果或是芒果，就算是胜利过关。而在2011年参加App高手争霸时的评审看过展示后，都不约而同地腼腆一笑，那场比赛是三大阵营的App共同比赛，光是触及敏感议题就不易出线。

而在2012年再度将该款游戏中的苹果改成电子发票图标，主题改为发票鬼集王"Invoice Runner"，参加中区税务局的App创意设计大赛获得佳作。

## 15-3 炸弹超人（Bomb King）

笔者开发的砖块系列，是继Lode Runner、Bricks Fighter之后的第三发。

原名为Bomber Man的炸弹超人，应该不用太多详细介绍的经典游戏。其实这款游戏与Lode Runner有着蛮大的相似度，有着一样的操控模式，虽然一个是纵切面的视角（Lode Runner），而一个是横切面（Bomb King），反而纵切面视角要处理从上而下的落下动作，横切面就简单许多。

Bomb King的游戏乐趣在于玩家要控制手上有限的炸弹数量，将炸弹放在正确的放置，才能炸出藏在砖块墙壁中的宝物，并且救出小鸡。详细完整原始码请参考本书所附的光盘。

整体游戏架构也与Lode Runner类似，放在此说明只是想要使读者明白一点，事实上许多游戏在实质上都是非常类似的处理。试回想，马里奥不也是在玩砖块吗？会跳，那是用当时行进的速度，加上其他因素来决定跳跃的力量而已，并且将关卡改为滚动条式处理；再进化就又可以改编魂斗罗游戏了……

Bomb King曾挤进Google play市场休闲游戏百大热门。

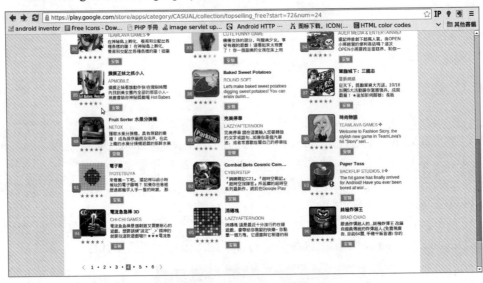

欢迎画面：

关卡选择画面处理。

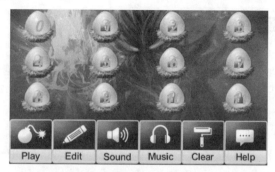

主要游戏画面：

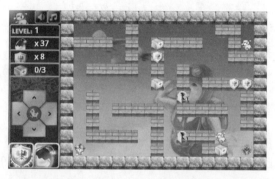

胜利过关画面：

游戏解说画面：

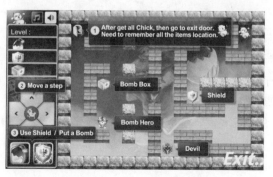

## 15-4 其他应用程序开发项目

上述三个小节都是以游戏为主轴的开发，而一般的文件或是相关书籍介绍，都会提及制作2D或是3D的游戏项目大多数是以SurfaceView来处理，为什么？作者Brad也不知道为何。但是Brad用实际的项目开发来提出反证，就是制作2D游戏使用自定义View即可达到功效，至少不是说说而已，是以实际项目开发证明，并且可以同时满足手机和大到10寸的平板电脑（至少Brad的TF-101变形金刚第一代预购版）都可以跑得非常顺畅。一个结论，在技术层面上，别人说的是我的参考，事实是自己证明出来的。专家，很多甚至是自己专门骗人家都还不自知的，因为说说建议而已可以不用负责任，这类的专家常见于各大媒体论坛场合，包括Brad本人，重点是学习者不应该只照单全收，而是需要马上做实验。

再来完成前面章节中的触控签名App吧。

### ■ 15-4-1 个性签名产生器

换名称了？总要有个响亮名称，才能衬托出有自信的App，所以从此改称为个性签名产生器。还记得开发状况吧，目前可以签名，有Clear/Undo/Redo的功能，继续开发下去的功能是：

- 改变颜色。
- 改变签字笔的粗细。
- 最后可以存盘。

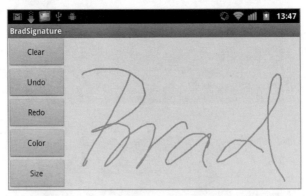

改变颜色对于目前程序而言，绝对是相当简单的一件事，只需要将Paint对象实体设定为用户想要的颜色即可，但是如何提供给用户一个友善的接口呢？

这边就先不考虑自行开发，先上网看看有没有适合的API可以使用。Brad的开发原则："站在巨人的肩膀上面，可以跑得比较快，前提是这位巨人愿意"。

上网找到Android Open Source Project的开放原始代码如下：

```
ColorPickerDialog.java
/*
 * Copyright (C) 2007 The Android Open Source Project
 *
 * Licensed under the Apache License, Version 2.0 (the "License");
 * you may not use this file except in compliance with the License.
 * You may obtain a copy of the License at
 *
 * http://www.apache.org/licenses/LICENSE-2.0
 *
 * Unless required by applicable law or agreed to in writing, software
 * distributed under the License is distributed on an "AS IS" BASIS,
 * WITHOUT WARRANTIES OR CONDITIONS OF ANY KIND, either express or implied.
 * See the License for the specific language governing permissions and
 * limitations under the License.
 */
package com.example.android.apis.graphics;

import android.R;
import android.os.Bundle;
import android.app.Dialog;
import android.content.Context;
import android.graphics.*;
import android.view.MotionEvent;
import android.view.View;

public class ColorPickerDialog extends Dialog {

    public interface OnColorChangedListener {
        void colorChanged(int color);
    }

    private OnColorChangedListener mListener;
    private int mInitialColor;

    private static class ColorPickerView extends View {
        private Paint mPaint;
        private Paint mCenterPaint;
        private final int[] mColors;
```

```java
            private OnColorChangedListener mListener;

        ColorPickerView(Context c, OnColorChangedListener l, int color) {
            super(c);
            mListener = l;
            mColors = new int[] {
             0xFFFF0000, 0xFFFF00FF, 0xFF0000FF, 0xFF00FFFF, 0xFF00FF00,
             0xFFFFFF00, 0xFFFF0000
            };
            Shader s = new SweepGradient(0, 0, mColors, null);

            mPaint = new Paint(Paint.ANTI_ALIAS_FLAG);
            mPaint.setShader(s);
            mPaint.setStyle(Paint.Style.STROKE);
            mPaint.setStrokeWidth(32);

            mCenterPaint = new Paint(Paint.ANTI_ALIAS_FLAG);
            mCenterPaint.setColor(color);
            mCenterPaint.setStrokeWidth(5);
        }

        private boolean mTrackingCenter;
        private boolean mHighlightCenter;

        @Override
        protected void onDraw(Canvas canvas) {
            float r = CENTER_X - mPaint.getStrokeWidth( )*0.5f;

            canvas.translate(CENTER_X, CENTER_X);

            canvas.drawOval(new RectF(-r, -r, r, r), mPaint);
            canvas.drawCircle(0, 0, CENTER_RADIUS, mCenterPaint);

            if (mTrackingCenter) {
                int c = mCenterPaint.getColor( );
                mCenterPaint.setStyle(Paint.Style.STROKE);

                if (mHighlightCenter) {
                    mCenterPaint.setAlpha(0xFF);
                } else {
                    mCenterPaint.setAlpha(0x80);
                }
                canvas.drawCircle(0, 0,
```

```
                        CENTER_RADIUS + mCenterPaint.getStrokeWidth( ),
                    mCenterPaint);

            mCenterPaint.setStyle(Paint.Style.FILL);
            mCenterPaint.setColor(c);
        }
    }

    @Override
    protected void onMeasure(int widthMeasureSpec, int heightMeasureSpec) {
        setMeasuredDimension(CENTER_X*2, CENTER_Y*2);
    }

    private static final int CENTER_X = 100;
    private static final int CENTER_Y = 100;
    private static final int CENTER_RADIUS = 32;

    private int floatToByte(float x) {
        int n = java.lang.Math.round(x);
        return n;
    }
    private int pinToByte(int n) {
        if (n < 0) {
            n = 0;
        } else if (n > 255) {
            n = 255;
        }
        return n;
    }

    private int ave(int s, int d, float p) {
        return s + java.lang.Math.round(p * (d - s));
    }

    private int interpColor(int colors[], float unit) {
        if (unit <= 0) {
            return colors[0];
        }
        if (unit >= 1) {
            return colors[colors.length - 1];
        }
```

```java
            float p = unit * (colors.length - 1);
            int i = (int)p;
            p -= i;
            // now p is just the fractional part [0...1) and i is the index
            int c0 = colors[i];
            int c1 = colors[i+1];
            int a = ave(Color.alpha(c0), Color.alpha(c1), p);
            int r = ave(Color.red(c0), Color.red(c1), p);
            int g = ave(Color.green(c0), Color.green(c1), p);
            int b = ave(Color.blue(c0), Color.blue(c1), p);

            return Color.argb(a, r, g, b);
        }
    private int rotateColor(int color, float rad) {
        float deg = rad * 180 / 3.1415927f;
        int r = Color.red(color);
        int g = Color.green(color);
        int b = Color.blue(color);

        ColorMatrix cm = new ColorMatrix( );
        ColorMatrix tmp = new ColorMatrix( );

        cm.setRGB2YUV( );
        tmp.setRotate(0, deg);
        cm.postConcat(tmp);
        tmp.setYUV2RGB( );
        cm.postConcat(tmp);

        final float[] a = cm.getArray( );

        int ir = floatToByte(a[0] * r +  a[1] * g +  a[2] * b);
        int ig = floatToByte(a[5] * r +  a[6] * g +  a[7] * b);
        int ib = floatToByte(a[10] * r + a[11] * g + a[12] * b);

        return Color.argb(Color.alpha(color), pinToByte(ir),
                          pinToByte(ig), pinToByte(ib));
    }
    private static final float PI = 3.1415926f;

    @Override
    public boolean onTouchEvent(MotionEvent event) {
```

```
        float x = event.getX( ) - CENTER_X;
        float y = event.getY( ) - CENTER_Y;
        boolean inCenter = java.lang.Math.sqrt(x*x + y*y) <= CENTER_RADIUS;

           switch (event.getAction( )) {
               case MotionEvent.ACTION_DOWN:
                   mTrackingCenter = inCenter;
                   if (inCenter) {
                       mHighlightCenter = true;
                       invalidate( );
                       break;
                   }
               case MotionEvent.ACTION_MOVE:
                   if (mTrackingCenter) {
                       if (mHighlightCenter != inCenter) {
                           mHighlightCenter = inCenter;
                           invalidate( );
                       }
                   } else {
                       float angle = (float)java.lang.Math.atan2(y, x);
                       // need to turn angle [-PI ... PI] into unit [0....1]
                       float unit = angle/(2*PI);
                       if (unit < 0) {
                           unit += 1;
                       }
                       mCenterPaint.setColor(interpColor(mColors, unit));
                       invalidate( );
                   }
                   break;
               case MotionEvent.ACTION_UP:
                   if (mTrackingCenter) {
                       if (inCenter) {
                         mListener.colorChanged(mCenterPaint.getColor( ));
                       }
                       mTrackingCenter = false; // so we draw w/o halo
                       invalidate( );
                   }
                   break;
           }
```

```
            return true;
        }
    }

    public ColorPickerDialog(Context context,
                            OnColorChangedListener listener,
                            int initialColor) {
        super(context);

        mListener = listener;
        mInitialColor = initialColor;
    }

    protected void onCreate(Bundle savedInstanceState) {
        super.onCreate(savedInstanceState);
        OnColorChangedListener l = new OnColorChangedListener( ) {
            public void colorChanged(int color) {
                mListener.colorChanged(color);
                dismiss( );
            }
        };
        setContentView(new ColorPickerView(getContext( ), l, mInitialColor));
        setTitle("Pick a Color");
    }
}
```

心中先感谢原创的作者，再好好检视一下程序代码，发现就是一个自定义Dialog，Dialog的重点是必须建构出对象实体之后调用show( )方法，并实作相关的监听事件方法即可，于是就直接拿来使用，改为MyColorPicker.java，放在相同的Package之下。

先在PaintView.java中建立两个方法：
- int getColor( )：取得目前的颜色值。
- void setColor(int newcolor)：设定新的颜色值。

```
int getColor( ){
    return paintLine.getColor( );
}

void setColor(int newcolor){
    paintLine.setColor(newcolor);
}
```

再回到 MainActivity.java 中，增加处理 chcolor 的按钮对象功能，就变成非常简单的一件事了。

```
chcolor = findViewById(R.id.chcolor);
chcolor.setOnClickListener(new OnClickListener( ) {
    @Override
    public void onClick(View v) {
    new MyColorPicker(MainActivity.this, new OnColorChangedListener( ) {
            @Override
            public void colorChanged(int color) {
                pview.setColor(color);
            }
        }, pview.getColor( )).show( );
    }
})
```

按下 Color 按钮后呈现如下图所示：

试着签名看看：

再来处理改变签字笔粗细及存盘功能，先将版面配置作些许改动：

```xml
<LinearLayout xmlns:android="http://schemas.android.com/apk/res/android"
    android:layout_width="match_parent"
    android:layout_height="match_parent"
    android:orientation="horizontal" >

    <LinearLayout
        android:layout_width="match_parent"
        android:layout_height="match_parent"
        android:layout_weight="4"
        android:background="#1100ff00"
        android:orientation="vertical" >

        <Button
            android:id="@+id/clear"
            android:layout_width="match_parent"
            android:layout_height="wrap_content"
            android:layout_weight="1"
            android:text="Clear" />

        <Button
            android:id="@+id/undo"
            android:layout_width="match_parent"
            android:layout_height="wrap_content"
            android:layout_weight="1"
            android:text="Undo" />

        <Button
            android:id="@+id/redo"
            android:layout_width="match_parent"
            android:layout_height="wrap_content"
            android:layout_weight="1"
            android:text="Redo" />

        <Button
            android:id="@+id/chcolor"
            android:layout_width="match_parent"
            android:layout_height="wrap_content"
            android:layout_weight="1"
            android:text="Color" />
```

```xml
<Button
    android:id="@+id/save"
    android:layout_width="match_parent"
    android:layout_height="wrap_content"
    android:layout_weight="1"
    android:text="Save" />

<SeekBar
    android:id="@+id/chsize"
    android:layout_width="match_parent"
    android:layout_height="wrap_content"
    android:layout_weight="1"
    android:text="Size" />
    </LinearLayout>

    <LinearLayout
        android:layout_width="match_parent"
        android:layout_height="match_parent"
        android:layout_weight="1"
        android:orientation="vertical" >

        <tw.brad.apps.BradSignature.PaintView
            android:id="@+id/pview"
            android:layout_width="match_parent"
            android:layout_height="match_parent"
            android:background="#44ffff00" />
    </LinearLayout>
</LinearLayout>
```

最后呈现如下：

先从调整签字笔粗细开始，因为是使用SeekBar，在之前单元的音乐播放上使用过，应该不陌生。先从PaintView下手处理两个方法：
- int getSize( )：传回目前粗细大小。
- void setSize( )：设定目前粗细大小。

```
int getSize( ) {
    return (int)paintLine.getStrokeWidth( );
}
void setSize(int newsize) {
    paintLine.setStrokeWidth(newsize);
}
```

回到MainActivity.java中：

```
chsize = (SeekBar)findViewById(R.id.chsize);
chsize.setMax(24);
chsize.setProgress(pview.getSize( ));
chsize.setOnSeekBarChangeListener(new OnSeekBarChangeListener( ) {
    @Override
    public void onStopTrackingTouch(SeekBar seekBar) {

    }

    @Override
    public void onStartTrackingTouch(SeekBar seekBar) {

    }

    @Override
    public void onProgressChanged(SeekBar seekBar, int progress,
            boolean fromUser) {
        if (fromUser){
            pview.setSize(progress);
        }
    }
});
```

这样就可以变更签字笔粗细。

再来就将签名文件存盘发送出去，建议将背景改为透明方式来处理签名图像文件。而要将View的显示内容进行存盘，应该事先调用setDrawing CacheEnabled(true)，否则预设为false。

```
pview = (PaintView) findViewById(R.id.pview);
pview.setBackgroundColor(Color.TRANSPARENT);
pview.setDrawingCacheEnabled(true);
```

处理按下 Save：

```
save = findViewById(R.id.save);
save.setOnClickListener(new OnClickListener( ) {
    @Override
    public void onClick(View v) {
        FileOutputStream fout;
        try {
            fout = new FileOutputStream("/mnt/sdcard/bradsign.png");
            boolean isSaveOK = pview.getDrawingCache( ).compress(
                        CompressFormat.PNG, 100, fout);
                if (isSaveOK) {
                    Intent i = new Intent(Intent.ACTION_SEND);
                    i.setType("message/rfc822");
                    i.putExtra(Intent.EXTRA_EMAIL,
                        new String[] { "brad@brad.tw" });
                    i.putExtra(Intent.EXTRA_SUBJECT, "Brad的签名");
                    i.putExtra(Intent.EXTRA_TEXT, "如附件");
                    i.putExtra(Intent.EXTRA_STREAM, Uri.fromFile(new File(
                            "/mnt/sdcard/bradsign.png")));
                    try {
                            startActivity(Intent.createChooser(i,
                                "Send mail..."));
                        } catch (android.content.
                            ActivityNotFoundException ex) {
                            Toast.makeText(MainActivity.this, "ERR",
                                Toast.LENGTH_SHORT).show( );
                    }
                }
        } catch (FileNotFoundException e) {
        }
    }
 })
```

要记得开启相关的权限：

```
<uses-permission android:name="android.permission.WRITE_EXTERNAL_STORAGE"/>
<uses-permission android:name="android.permission.INTERNET"/>
```

这样完成一个小品项目。

### ■ 15-4-2 开发观念原则

目前许多在Android App的项目开发中，其实架构流程并不大，比起以往Web-based的项目而言，真的非常小，真要说团队合作，大概顶多就是两三个人就可以，一个开发程序、一个视觉艺术，加上一个音效制作。但是Brad经常看到许多创投育成中心，不断强调团队开发合作模式的重要性，还必须要特定地点大家非得面对面讨论，激发创意灵感等。Brad却深深不以为然，至少对于程序开发的部分真的没有那么复杂，就是做自己想做的东西，做出来的东西都无法满足自己的要求，更何况是全世界用户的严酷考验。面对面互相讨论的事情，难道视频无法做到吗？而在台中、花莲、台东所Google搜寻出来的结果会与在台北的结果不一样吗？

因此，一个重要的观念就是，Android Apps给了许多独立开发者一个很大的开发舞台。学习者老是认为开发工作有多艰巨，事实上就是去做而已（Just do it），以本书中所开发的个性签名，重点就是如何规划以数据结构来处理签名划线，如果处理得宜，后面的功能性开发就非常容易。通常这只是授课实务范例而已。

# 16 Chapter

## 第16课　App发布

16-1　包装发布到 Google Play

16-2　App创意开发与比赛经验心得分享

# 16-1 包装发布到Google Play

Google Play算是在Android Apps上非常大的交易市场。开发者可以通过这个平台，将所开发的Android App上传上去后，以付费或是免费的机制散布到世界上大多数地区的用户，而用户得以利用这个平台机制，寻找喜爱的App，进而免费或是付费下载使用。

## ■ 16-1-1 包装成为APK

开发完成的项目，终于要与世人见面了！以下就以Bricks Fighter项目实际处理方式介绍认识。

开启项目目录下的AndroidManifest.xml中的Manifest的页签，看到有两项重要数据：

- Version code：版本码。
- Version name：版本名称。

• Version code是开发者识别专用的码，每次上传到Google Play的时候，会检查该码是否比上次上传的数字要大，笔者建议的编码方式是YYYYMMDD01，也即公元年＋月份＋日期＋两位流水码，因为今天的日期永远比昨天大，流水码配置两位数，一天要能更新发布99次应该是不常见的事，而往后还可以通过该码知道上次发布的时间。

Version name则为用户会看到的版本名称。

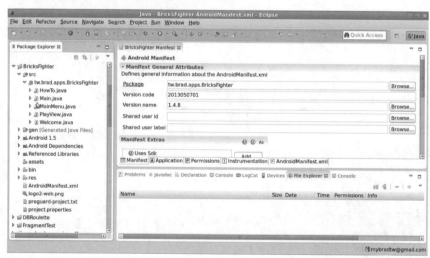

接下来点按项目名称右键，"Android Tools"→"Export Signed Application Package"，如下展开：

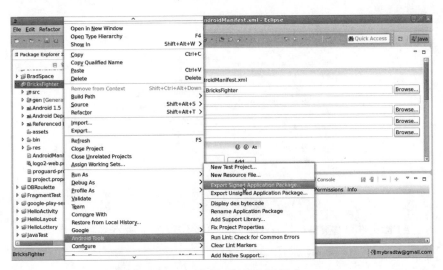

开始进行一连串询问精灵窗口，选择项目进行导出。

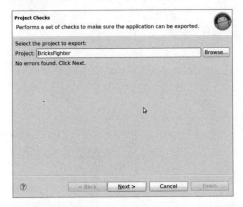

第一次发布没有任何keystore文件，必须建立一组。而该App往后的更新发布都是以该组keystore进行，因此务必善加保存该文件。

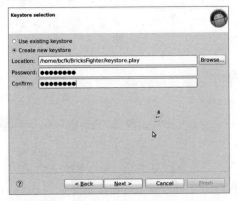

填写该组keystore相关数据，以下图举例为最基本要填写的项目（有效期，官方建议是25年）。

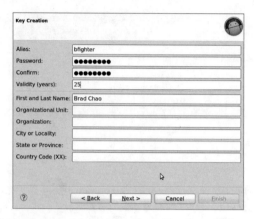

设定APK文件产生的目录及文件名,笔者建议加上版本名称以作识别。

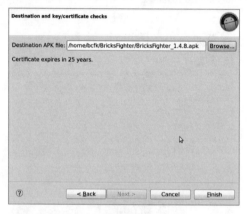

按下"Finish"后开始进行。

## 16-1-2　首次注册开发者

开发者想要通过Google Play交易平台发布,有以下几个重点要进行:
① 注册Google账号,最简单的方式就是Gmail的账号即可。
② 注册Google Play开发人员控制台账号。
③ 25美金。
④ 注册程序依照官方网站说明为48小时内,笔者实际经验应该是24小时之内(甚至更快)。
⑤ 如果要贩卖付费Apps,之前需要另外注册Google Checkout商家,从2013年3月起,已经由Google电子钱包商家中心取代。

笔者对于台湾地区许多开发者的建议是,国际市场应该比台湾地区的消费市场大很多,针对创意性的创作,不要只是着眼开发应用层面较为区域性的Apps。当然,如果是一家中小企业或是机关组织为了其商业或是宣传便民之目的而开发的作品,那就另当别论。有公司要开发小区管理系统,仅提供该公司管辖的小区住户使用的Apps，功能包含挂号信通知、公告通知等。有特殊目的需求，当然就开发，那叫做"接案子"。如果现在闲下来，想要进行创意性的开发，笔者不会想到要做"某地美食报你知"这类的Apps，因为从下载广告或是付费用户数量观点来看，付出的开发代价应该不太值得的。

注册程序中的Google Play开发人员注册非常简单：
① 使用Google账户登入。
② 接受开发人员协议。
③ 支付注册费。
④ 填妥账户详细数据。

搜寻play developer console即可找到开发人员控制台。当以Google账户登入后，会看到以下网页内容：

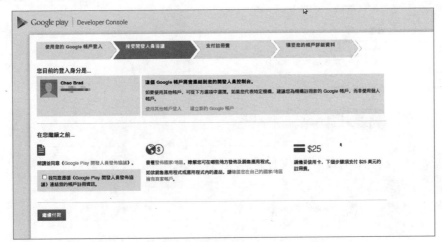

跟着步骤进行就应该可以顺利完成注册（重点是要付25美金）。应该会收到如下图的通知信件：

从此之后，如果都是以该账号发布 Apps，都不再需要支付其他费用，这一点与其他两个阵营的机制是不一样的。

### 16-1-3　发布APK到Google Play

App 已经开发完毕，也包装成为 .apk；而 Google Play 也已经注册为开发者，就开始进行发布的动作了。

首先登录到开发者控制台，直接点按"新增应用程序"后，出现如下画面：

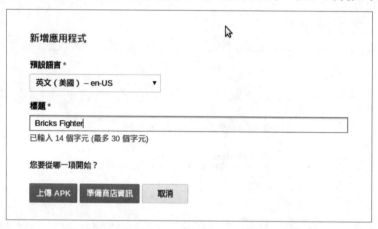

开始将已经包装好的 APK 文件进行上传。

正常上传中：

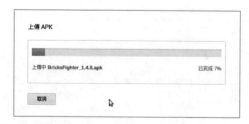

完成上传后出现的状况报告：

点选左侧的商品信息后填写相关数据：

屏幕截图的提供有助于用户了解App的外观。

填完以上数据后，点按"定价与发布"即可开始设定付费与免费相关机制。
一切顺利后就算是完成上架程序了。

## 16-2  App创意开发与比赛经验心得分享

还是一句老话，笔者Brad并非专家（也不是专门骗人家，从事教育工作时戏谑为误人子弟），仅就这近四年来玩Android Apps开发及教学经验作分享。而本书中所分享的开放原始代码，也提供最完整的项目程序代码，希望读者可以检视到最完整的应用，但是Brad并非最专业开发工作者，所以提供的原始代码也并非专业的开发考虑，期待读者能以学习的角度参考使用。

创意的产生是开发者面临的第一个问题，Brad也不是创意达人，只想做自己想要的东西而已。2010年间在"资策会"辅导学员制作专题时，曾经提供学员一个想法："既然移动装置都有因特网的应用，何不写个类似MSN的App，这样用户就可以省下短信的费用"。后来学员还是做了其他主题，而笔者当年如果完成这个想法，那现在可能大家不是用Line，而是BradMSN……（后悔中）。无论如何，笔者还是做自己想做的东西，前年开始想写些红白机系列的游戏，于是就从Lode Runner下手，陆续有炸弹超人及打砖块等。这样的过程至少满足自己所需，而且也和儿子玩得非常开心。当然，如果也能带给全世界不相识的玩家快乐，那又是不同的喜悦……

这是一位来自瑞典的玩家，不断要求增加关卡……

笔者增加设计更多关卡，但是似乎仍无法满足……

最后只好承诺会在圣诞节前推出新关卡，目前的160关都是被这位玩家给逼出来的。

这又是另外一位来自德国的玩家，帮我设计关卡，非常热心。

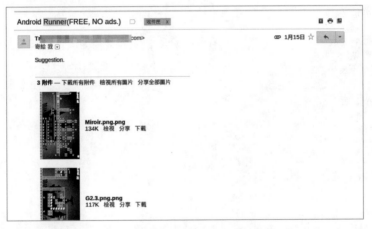

这位玩家真的玩不出关卡，来要求救秘诀，我只好拍成youtube给他参考。

太多来自世界各地玩家享受这款游戏，即使是免费，至少人生中有这样的际遇，就让Brad觉得相当值得，不是用金钱可以换来的价值。

而每年台湾地区都有许多相关的比赛可以参加。Brad本来不想参与其中，但是曾经有位学员半开玩笑地说："老师，你自己也没有参加过比赛呀！"就在出版一本Android书期间，最直接的示范动作，就是将授课范例，参加当年2010年间的"创意成金"活动，共有四个范例参加，被选中了三个范例：《猜数字游戏》、《我的待办事项》及《旅游全纪录》，也得到一笔奖金。当年原本是将猜数字游戏范例放进上一本书中介绍，基于活动规则，还特定将公开发表的书中范例变更为猜数字大小游戏，而不是1A2B的猜数字游戏。如此一来，学生就应该比较能够信服，所以后来陆续参加的比赛，都是为了增加授课及辅导的实战经验。

参加比赛可以督促开发者将一个作品进行完整化的包装，而不是一天拖过一天的开发。

最重要的是，参加比赛千万不要得失心太重，参赛的过程与经验的积累，才有更重要的实质意义。